Advances in Intelligent Systems and Computing

Volume 1338

AF863170

The series “Advances in Intelligent Systems and Computing” contains publications on theory, applications, and design methods of Intelligent Systems and Intelligent Computing. Virtually all disciplines such as engineering, natural sciences, computer and information science, ICT, economics, business, e-commerce, environment, healthcare, life science are covered. The list of topics spans all the areas of modern intelligent systems and computing such as: computational intelligence, soft computing including neural networks, fuzzy systems, evolutionary computing and the fusion of these paradigms, social intelligence, ambient intelligence, computational neuroscience, artificial life, virtual worlds and society, cognitive science and systems, Perception and Vision, DNA and immune based systems, self-organizing and adaptive systems, e-Learning and teaching, human-centered and human-centric computing, recommender systems, intelligent control, robotics and mechatronics including human-machine teaming, knowledge-based paradigms, learning paradigms, machine ethics, intelligent data analysis, knowledge management, intelligent agents, intelligent decision making and support, intelligent network security, trust management, interactive entertainment, Web intelligence and multimedia.

The publications within “Advances in Intelligent Systems and Computing” are primarily proceedings of important conferences, symposia and congresses. They cover significant recent developments in the field, both of a foundational and applicable character. An important characteristic feature of the series is the short publication time and world-wide distribution. This permits a rapid and broad dissemination of research results.

Indexed by SCOPUS, DBLP, EI Compendex, INSPEC, WTI Frankfurt eG, zbMATH, Japanese Science and Technology Agency (JST), SCImago.

All books published in the series are submitted for consideration in Web of Science.

More information about this series at http://www.springer.com/series/11156

Valentina Colla · Costanzo Pietrosanti
Editors

Impact and Opportunities of Artificial Intelligence Techniques in the Steel Industry

Ongoing Applications, Perspectives and Future Trends

Springer

Editors
Valentina Colla
TeCIP Institute
Scuola Superiore Sant'Anna
Pisa, Italy

Costanzo Pietrosanti
Danieli Automation
Buttrio, Italy

ISSN 2194-5357 ISSN 2194-5365 (electronic)
Advances in Intelligent Systems and Computing
ISBN 978-3-030-69366-4 ISBN 978-3-030-69367-1 (eBook)
https://doi.org/10.1007/978-3-030-69367-1

This Springer imprint is published by the registered company Springer Nature Switzerland AG
The registered company address is: Gewerbestrasse 11, 6330 Cham, Switzerland

Preface

Within the digital factory of the future, a huge volume of heterogeneous and often non-structured data is collected from the different stages of the production chain as well as from external sources. Such massive generation of data enables the efforts toward the extended *sensorization* of steel plants by continuously monitoring all the aspects of manufacturing processes and relevant assets. Moreover, the improvement and the development of existing and new sensors systems, capable of high sampling rates to target the real-time applications, give a substantial contribution to such trends.

Therefore, artificial intelligence (AI) and machine learning (ML) play a key role in the full exploitation of the potential value hidden into such huge volume of data. This trend becomes more and more evident looking to the increasing demand of "intelligence" of autonomous machines and systems to mimic aspects of the behavior of human beings.

Today, in the steel context as well, AI and ML are becoming important pillars of the human–machine cooperative environment within smart networks. Moreover, AI and ML enable the industrial transition from *reactivity* to *predictivity* making systems faster and effective as never seen before. Look at the change in maintenance, to event forecasting, to optimized recovery strategies, and they also support *preventatively* organizations discovering common aspects describing event and, often, accident dynamics.

Consequently, the steel sector is addressing important parts of the global figure invested in the digital transformation of the manufacturing and supply chains. This is done in view of facing the challenge of big data exploitation, by considering intensive application of AI and ML as a major enabler for improving its competitiveness on the global market and promoting the sustainable growth.

In this context, the idea to dedicate a specific workshop to the application of AI and ML in the steel sector emerged to the aim of giving a contribution to the state of the art of the diffusion of such technologies, on one side, and defining the key directions of research, development and innovation responding to demand of the steel industry.

Concerning the state of the art, AI and ML are not new in the steel sector being thoroughly accepted techniques in the wave of the Industry 4.0.

From the industrial standpoint, several AI and ML applications are already implemented or under development in the steel factories. Many applications are focused on process optimization, quality enhancement, yield increase and safety improvement with the ambition to accelerate the process of making steel manufacturing sustainable according to the mandate of the EU.

Another key aspect is represented by the already mentioned ambition of mimicking humans that largely contribute to decrease risks in harsh operations in the steel plants. The potential impact of efficiency and safety enhances process control and dramatically improves situation awareness, by improving workers' health and safety conditions while reducing physical and mental stresses.

For all such reasons, the Smart Factory Focus Group of the European Steel technology Platform (ESTEP) matured the idea to dedicate a specific event aimed at analyzing the state of the art, the enabling factors and the barriers together with consequences and opportunities offered by AI and ML in the steel sector. In addition, there was the need to rationalize the fascination of creating digital and cyber-physical systems, which are capable of learning from experience through ML with the ambition of maintaining and strengthening the centrality of human role.

The abovementioned idea became concrete in 2020, when ESTEP organized the workshop entitled *"Impact and opportunities of Artificial Intelligence in the Steel Industry."* Such workshop was aimed at encouraging the participation of the steel community and of other industrial sectors to share perceptions, experiences, needs and expectations and fears concerning AI and ML as well as at providing indications for future targeted research, development and innovation actions. Societal aspects and deep consideration of the impact on workforce and education systems have been also considered as major issues in empowering human role in the smart factory. However, to succeed in this idea, new skills are necessary and new relationships, organizational models and tools must be established and implemented between demand and offer of such new skills, to foster a conscious and fearless approach to AI and ML-based tools and techniques, by seizing their opportunities and potential benefits.

The selected venue for the workshop was Pisa (Italy), as the workshop was hosted by Scuola Superiore Sant'Anna. Unfortunately, due to Covid-19 outbreak and consequent travel and meetings restrictions, the event was finally organized online in the form of four weekly sessions which were established by ESTEP. Nonetheless, the workshop has been characterized by the relevant participation of experts, industrial and ICT engineers and managers, solution providers and system integrators, suppliers, universities and research institutes, consulting companies, national and international organizations operating in the steel field as well as in other industrial sectors.

Contributions and presentations were encouraged mostly (but not exclusively) related to the following topics, which are considered relevant with respect to the steel context:

- industrial needs and research and development directions in AI and ML;
- data management and data quality issues relevant for the application of AI and ML;
- advantages, capabilities and trends in the industrial application of AI and ML;
- constraints, opportunities and threats of data-driven technologies, including cyber-security aspects;
- AI standards and management models in process industry;
- impact of AI and ML: workforce, operations, systems and organization;
- examples and experiences of implementation of AI- and ML-based applications in the steel sector.

As a result, the most noticeable outcomes observed in the workshop are summarized in the following.

AI and ML are not new and are now widely used in the steel context for applications not only in the plant of the champions of the Industry 4.0 concept that not necessarily are big steel manufacturers.

AI and ML are perceived by the users and sustained by the experts as one of the active gamechanger within the evolution of the digital transformation.

AI applications are devoted today mainly to process optimization, yield improvement, quality management and reduction of consumptions. Few applications are running in the context of environmental challenge; however, they are increasing.

Despite all, front edge applications are not yet frequent. However, the deployment of AI and ML is seen to increase rapidly also for integrated multi-objective optimization and complex event management.

The role of data and data technological landscape as enablers of the AI success is broadly clear; however, some players run ICT architectures still based on the ISA 95 standard. This is due to the large investments to change the ICT architecture that are some order of magnitude higher than pilots on AI. Despite that, a sort of inversion of priorities pushes users to invest before in the AI/ML application than on the modernization of the ICT architecture. This could open in the medium terms a gap between the expected benefits and the real possibility to the full exploitation of the potential.

Two factors related to people emerged: the necessity of new skills empowered by new ways and tools for advanced training and the capability to attract talents in the steel industry. Despite the ever-increasing attention on the human factor, it is necessary to create rapidly new skills in presence of the unbalanced demand and offer of new skills. In consequence, it is necessary to promote the appeal of steel careers and, more in general, the image of the steel plants as high-tech sites.

In conclusion, a lot of efforts and developments are still possible and research on fundamental aspects of mathematics, logic and informatic sciences must be promoted creating the conditions for facilitating the cooperation between public and private players. This is foreseen in the instruments and the opportunities offered by the incoming Horizon-Europe framework program. There, digitalization is declared as one of the key development factors to support the social and industrial model of Europe.

This book collects a selection of the most interested interventions, which were proposed in the abovementioned workshop, which was made after the conference, based also on the conference chairs assessment quality of presentation and audience interest.

December 2020

Valentina Colla
Costanzo Pietrosanti

Organization

Workshop Chairs

Valentina Colla	Scuola Superiore Sant'Anna, TeCIP Institute, Pisa, Italy
Costanzo Pietrosanti	Danieli Automation S.p.A., Buttrio, Italy

Workshop Board

Valentina Colla	Scuola Superiore Sant'Anna, TeCIP Institute, Pisa, Italy
Jerker Delsig	Luleå University of Technology, Luleå, Sweden
Stephane Mouton	Centre of Excellence in Information and Communication Technologies, Charleroi, Belgium
Marcus J. Neuer	Betriebsforschungsinstitut, Düsseldorf, Germany
Klaus Peters	European Steel Technology Platform, Brussels, Belgium
Costanzo Pietrosanti	Danieli Automation S.p.A., Buttrio, Italy

Contents

About the Authors

Dr. Eng. Valentina Colla obtained her Master Degree in Engineering from the University of Pisa in 1994 and her Ph.D. in Industrial and Information Engineering from Scuola Superiore Sant'Anna in 1998. At Scuola Superiore Sant'Anna, she became post-doctoral research fellow in 1998 and researcher in 2000. Since 2008, she is Technical Research Manager at Scuola Superiore Sant'Anna, where she currently works. In 2001, she funded a research team named "Steel and Industrial Automation Division," which became in 2016 the "Center of ICT for Complex Industrial Systems and Processes" (ICT-COISP). She is currently working within the Institute of Communication, Information and Perception Technologies (TeCIP) of Scuola Superiore Sant'Anna and is coordinator of ICT-COISP, whose activity is focused on process automation, modeling simulation and control of industrial processes, especially process and manufacturing industry. The research interests of her concern the application of statistical tools and AI-based techniques for data analytics, modeling, simulation, monitoring and control of industrial processes, with a particular focus on process and manufacturing industry, in particular within the iron and steel sector. She started developing research activity within the steel sector in 1996, working on the application of machine learning-based models for the prediction of mechanical properties of steel products. Since 2000, she participated to 66 EU-funded projects (8 as project coordinator) and in many projects funded by private companies. She is also an active member of the European Steel Technology Platform (ESTEP), where she is Co-chairman of the Focus Group "Circular Economy" and member of the Focus Group "Smart Factory." She is also an expert member of the Technical Group TGA5 "Steel factories – smart and human" of the Research Fund for Coal and Steel. She had several experiences as evaluator of proposals submitted in response to many European calls. She is Author and Co-author of about 120 papers published in national and international journals and about 220 papers presented in national and international conferences.

Dr. Eng. Costanzo Pietrosanti obtained his Master Degree cum Laude in Mechanical Engineering in July 1978 at the University of Rome La Sapienza. After one year of Assistant Researcher in Machine Design, he joined the Industrial Researcher at Centro Sperimentale Metallurgico SpA (CSM) in Rome, the Corporate Research Center of the Finsider Group, the public holding of the Italian steel manufacturing industry.

After three years spent in several industrial research projects both private and supported by public funds in the field of Mechanical Testing and Fracture Mechanics, he joined the former Snia BPD Group (now Avio firm) working for three years in the integration of the Ariane V rocket with the role of responsible of structural analysis.

In 1984, he came back to the Centro Sviluppo Materiali SpA (the evolution of the former Centro Sperimentale Metallurgico SpA) transformed into a supporting entity of the newly born public steel holding Ilva, active on the industrial market of consultancy and services for the Italian and European industry. He was the head of the Technical Computing Center implementing large computing facilities and creating the excellence center on nonlinear computing for material forming and process design participating to many private and public R&D projects in the field of the development and optimization of steel application in the automotive and oil and gas sectors. From 1989 and 1992, he was responsible for the installation of the first Italian plant of thin slab casting at the Acciai Speciali Terni site.

In 1992, he became Manager and Assistant to the CEO of the Ilva Group with the responsibility of the creation of the international R&D scouting and cooperation network.

In 1996, following the transformation of CSM into a private shareholding company, he was Director of the Automation and Process Control Division participating to many RFCS supported and private industrial projects. He retired in 2013 embracing the activity of Industrial Consultancy for the Danieli Automation SpA belonging to the Italian Automation and Plant Supply firm. From 2007 to 2019, he represented CSM in the ESTEP Working Group named Intelligent Integrated Manufacturing (I2M). Since 2019, he is Chairman of the group, currently named Smart Factory Focus Group. The scientific work is summarized through more than 50 publications in the field of advanced computing in mechanical analysis and design, Fatigue and Fracture Mechanics and Process Technology in the field of Industry 4.0.

Challenges and Frontiers in Implementing Artificial Intelligence in Process Industry

Marcus J. Neuer(✉), Andreas Wolff, and Norbert Holzknecht

Betriebsforschungsinstitut (BFI), Sohnstr. 65, 40237 Duesseldorf, Germany
marcus.neuer@bfi.de,
http://www.bfi.de

Abstract. The implementation of artificial intelligence faces different challenges of infrastructural, data related, security related and social scope. These aspects are discussed, reflecting on the requirements of introducing such a technology in a broader way. Machine learning, as subfield of artificial intelligence, benefits from advances in Big Data science. One example is the λ-architecture which can be used for the treatment of streamed process data in industrial applications. Digital twins are shown as further tool providing object-oriented data for machine learning applications. Yet, the increasing freedom of data transfer within a plant, as propagated by Industry 4.0, poses new risks for information technology and automation systems: security of those components is one of the big challenges. Here, artificial intelligence can seen as both risk and solution. A last relevant challenge is acceptance among the staff, as artificial intelligence is associated with fears. Counterstrategies for those fears are presented as a proposed guideline for real applications. Finally, current frontiers at process industry are considered and discussed. These include the need for strengthening the use of high-dimensional data availability, increased roll-out of optimisation concepts and rigorous progresses in semantic modelling of processes and process chains, in order to fully exploit the beneficial scope of artificial intelligence in industry.

Keywords: Machine learning · Artificial intelligence · Semantics · Social aspects · Big data · Industry 4.0 · IT and OT security

1 Introduction

The word data first appeared in the fifteenth century. At that time, proper planning of the fields, their efficient use and, of course, maximising the usable harvest played a crucial role for people. Since then, data was acquired, evaluated, interpreted and decisions were based on it. The fact that data provided insight into certain aspects, allowed different perspectives and made things visible that were not seen before, substantially changed the decision process and helped to steer and control processes. Not much changed since then, regarding the main idea behind taking data.

Today we are in the midst of a digital transformation, a disruption that is transforming our social life and the way we work. Digital process data has become

V. Colla and C. Pietrosanti (Eds.): ESTEP 2020, AISC 1338, pp. 1–12, 2021.
https://doi.org/10.1007/978-3-030-69367-1_1

widely available as a result of previous technological developments provided by the Big Data phase and the Industry 4.0 era. With this data availability, the next step is right in front of us: How do we use the data for something useful?

One answer to this question is machine learning (ML), or artificial intelligence (AI) in general, as it is able to extract information from data by learning relationships from them. ML and AI are two terms that are often, albeit erroneously, used synonymously. For a better understanding we should therefore consider the following quotations:

> "Artificial intelligence is the science and engineering of making computers behave in ways that, until recently, we thought required human intelligence" - Andrew Moore, Carnegie Melon University

> "Machine learning is the study of computer algorithms that allow computer programs to automatically improve through experience" - Tom Mitchell, Carnegie Melon University

From these quotes we see that machine learning is a subfield of artificial intelligence, namely the subfield that allows computers to learn something. To learn, these algorithms need data and out of this reason, ML was the field that quickly appeared promising for data evaluation. The term AI has much wider scope, including concepts like self-awareness, automation of decision processes, solutions to triage problems or even emotional interpretation.

In industry, most applications we currently see revolve around ML and stay within this subfield of AI. Beneath its recent story, it took nearly over a whole century, with several historic peak times of ML to reach its current substantial success. Indeed, many ground-braking research results were obtained eighty to twenty years ago, and ML saw quite a number of renaissance phases, the strongest clearly driven by the Big Data trend, more or less initiating what we call the era of artificial intelligence by now.

2 Challenges for AI in Process Industry

2.1 Technological and Infrastructural Challenges

European process industry operates plants at very different digitalisation levels. From the IT perspective, this landscape is a brown-field, because machinery has grown and changed over decades, enhancing older aggregates as well, as extending the equipment with new machinery.

Infrastructure is also one of the primary cost factors. Scalable technology for hosting modern data solutions requires substantial investments. It can be safely assumed that this is a serious barrier for any larger role-out of artificial intelligence applications. Many successful demonstrations of AI we have seen by now were financed via funded research, which also helped industrial users to accomodate with the technology and to experiment with it, in order to verify its expected economic impacts.

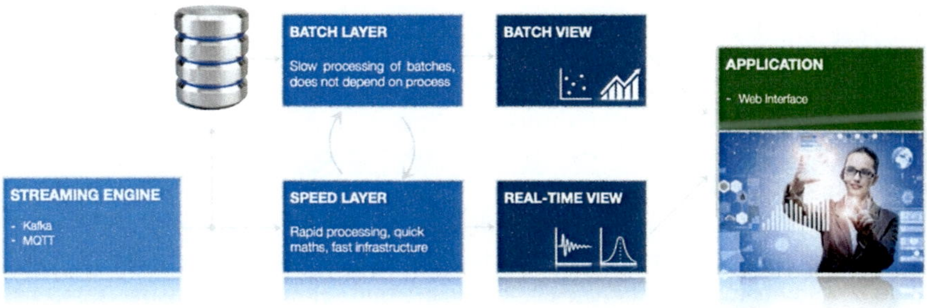

Fig. 1. Lambda-architecture for process data.

Flexibility remains key for the success. The solution to this challenge can therefore be formulated in the following way: find a hardware platform that is a) extensible to ensure future flexibility, b) serviceable inside the company, to change the operating system or software components, c) scalable, so that it is easy to add additional hardware for increasing performance, and d) secure, to not allow any attacks from outside (refer also to Sect. 2.3).

The remaining question is then to decide whether this infrastructure is on premise using so-called edge computing or in the cloud. Up to now, both approaches have been seen in industry. Those opting for the edge-based solution have in mind to stay in possession of their data all the time, a position dominating especially conservative industries like steel production. The contrary idea, to put all data into a managed cloud system appears on first sight more insecure. Nevertheless, any on premise activity requires that a company maintains a highly skilled IT team familiarised with the defense of threats and that should better be superior to the IT security specialists of cloud providers. Another aspect that supports the use of the cloud, is the outsourcing of risk. In case of proper legal agreements, data loss or harmful events can become the responsibility of the provider.

Marz and Warren describe in [1] details of the λ-architecture, a way to split data streams from processes into a slow processing branch, the so-called batch layer and a rapid processing branch, the speed layer. It is a paradigm from managing Big Data. Figure 1 yields these two branches in some more detail. The term λ originates from the form itself - the two branches can be regarded as the two legs of the letter λ. For process industry it is important that such an architecture (or similar) can be realised by the given infrastructure.

Once the appropriate structure is in place, machine learning techniques can be conveniently trained inside the batch layer of the λ-architecture. Slow training procedures can be executed, parsing historic and current data. After training the finalized models can be used for predictions within the speed layer, because they are quick with regard to execution time, much quicker than any training.

2.2 Data as Primary Resource that Drives Artificial Intelligence

With the rising amount of available process data being one of the primary reasons to need machine based data analytics, let us now shortly review some important aspects of this primary resource for the application of AI. In the previous section we saw constraints for the infrastructure that is needed to appropriately handle large amounts of process data. Big Data is characterised by a set of v's that refer to a) a high volume, b) big veracity, c) high velocity and d) has a great variety, as well, as e) the business value, the economic estimate how much business potential is contained in the data. A deeper treatise on what Big Data actually is can be found in [2].

At a certain point in time, data technologies and their rigorous business oriented exploitation were focused by several consultation firms and Big Data as a disruptive technology was soon identified to bring global change to modern economics. McKinsey devoted regular strategic reports on the topic covering e.g. chances [3], potential impact [4] and associated business needs. PricewaterhouseCoopers (PwC) put a similar focus on the topic in [5]. The hype around Big Data also spawned a plethora of different new software solutions which instantly became highly relevant for process data. Many production companies upgraded or are still upgrading their computational equipment to hold up with these latest efforts.

Why were these developments so important for the application of AI? Because for a long time, many data analytics approaches considered only highly aggregated quality features of products. The much higher resolved time-series or length-series of process data, not to mention any additional enrichment by mathematical transformations of these, have rarely been taken into account. In [6] one exemplary data reference architecture was presented, that integrated all process data and allowed to correlate these data to quality labels like the number of defects. Such infrastructures are needed to foster the application of artificial intelligence in the industrial environment.

Gziello et al. presented in [7] an example about a quality control system to improve the continuous casting process, where they significantly extended the number of technical parameters that went into their considerations. Similarly, Konrad et al. studied in [8] how to apply means of data analytics for quality control, surveilling selected real-time variables of a continuous rolling mill. Both works are just examples of a whole family of activities that took place even before the eve of Big Data but already pointed towards the necessity of increasing the number of variables. The acquisition of models in Big Data volumes is often done via distributed machine learning [9,10], and commonly involving strategies like deep learning [11] or ensemble algorithms. Today, works like Gao et al. [12] yield promising results in applying autoregressive networks directly on these high-dimensional raw data streams.

When considering the technical infrastructure as the hardware perspective, subsequent data management can be regarded as the software. It is a separate challenge how to handle the data in a suited way. Cyber-physical systems (CPS) are currently one of the most successful approaches for real-time data evaluation.

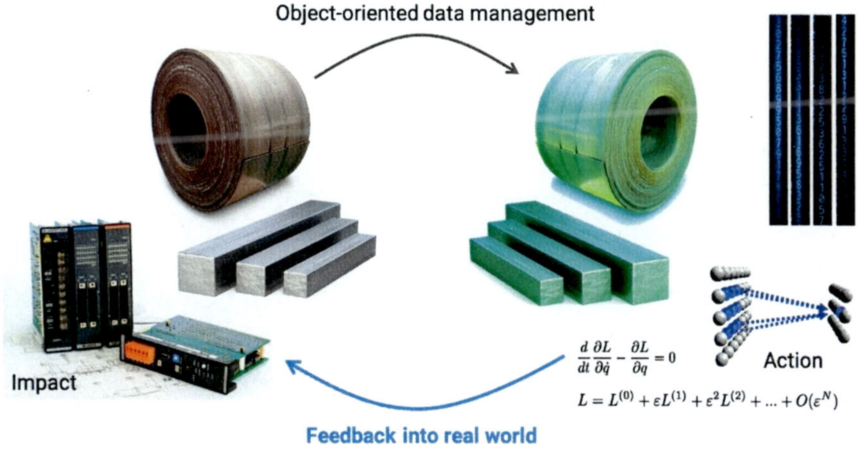

Fig. 2. Digital twins as ideal infrastructure for training machine learning concepts and for evaluating the trained models within the process chain.

Such systems use digital twin (DT) concepts to represent the cyber-aspect of the CPS. They require specific object-oriented database paradigms such as NoSQL. The latter offers data structures that can host unstructured and strongly heterogeneous types of information. Modern NoSQL databases are scalable and can be extended by new real or virtual instances. Such DBs are the backbone for putting digital twins into practise.

Figure 2 shows an illustration of the digital twin concept. Per definition, we presume a DT to be the digital representation of some real counterpart, which can be a product like depicted in the figure or a machine, or even a machinery component. A central notion of DT is that they can evaluate models in real-time, accompanying their real counterpart with forecasts, and use the model output to interact with the production chain: DT provide actively a feedback into the real-world. Potential model evaluations are here indicated by Lagrangian equations symbolising the analytical, first-principle approach and by neural network, as representation of AI modelling.

The benefit of using digital twins for the implementation of AI in process industry is twofold: a) they provide an easy access to data and labels for the training of AI models and b) it is later on the ideal place to evaluate these models for process forecasting.

2.3 Security Challenges

Industry 4.0 revolves around the broad vision of vertical and horizontal integration. This integration is a necessary precondition for the application of artificial intelligence, as it gathers the relevant data from different aggregates along the process chain. Unfortunately, the increasing degree of integration does not only have advantages: malicious codes, for instance from external attackers can exploit

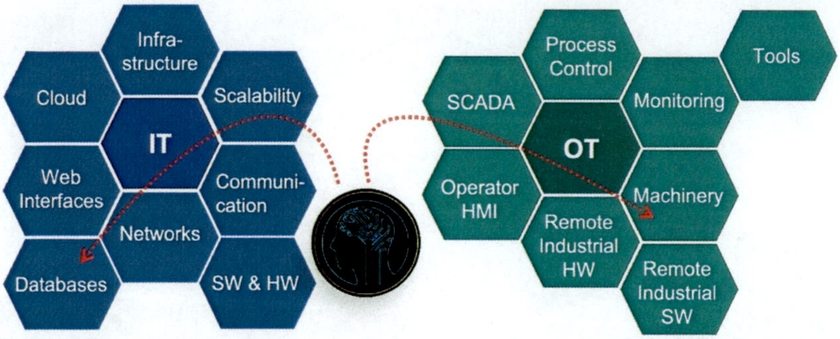

Fig. 3. AI solutions will require access to both, the information technology (IT) and the operational technology (OT) layers.

those network connections. Figure 3 shows some typical components of IT and OT systems.

This threat is even increased by the fact, that cybersecurity in production companies is often the aspect of information technology (IT). Operation technology (OT) components, such as the automation systems of the productions processes, are similarly vulnerable to attacks. After the success of the Industry 4.0 activities, they have become much more exposed.

Following types of attacks exists:

- **Ransomware attacks** infest and jam computers of a company, forcing the victims to deliver transfer some ransom to the attackers. This type of attack predominantly affects the IT layer.
- **Destructive attacks** aimed at important production processes (like furnaces) have been reported in the past. Aggressive malicious codes are intentionally injected into the automation systems, either copied via the network or simply manually transferred from a USB memory, confuse the controllers and drive the targeted process into a dangerous, damaging state.
- A third, but nonetheless highly important class of security threats are **non-destructive, latent attacks** that are not detected by the plant operators. These attacks just perform random perturbations on the processes, which remain hidden, because the attack also mimics normal control states.

Artificial intelligence is not only part of the problem here, it is essentially also part of a solution. In [13], Xu et al. show how variational autoencoders can be used to effectively perform an intrusion detection. Neuer et al. show in [14] that autoencoders can also be used for anomaly detection in automation systems, revealing the vulnerability of the current OT components and the strong needs to actively develop new strategies and software for attack detection.

2.4 Social Challenges Caused by Staff Acceptance

Introducing AI in broad scope for the process industries has also a strong social dimension and it brings along a whole set of issues. But it should be thoroughly underlined and understood, that these issues are not alone associated with AI, they are a barrier for any complex, algorithmic solution deployed in industry. Even mathematical formulas derived from first-principles may induce scepticism and reservation among workers.

As their motivations are not the same, we have to distinguish between the operating personnel and process experts. Staff that directly works with the machines might exhibit several fears, that contribute to their opinion. Following the illustrative picture in Fig. 3, we compiled in Table 1 a list of potential worker fears and proposed solutions for improving the acceptance.

The development of an onboarding strategy for AI, as also pointed out in the table, is of predominant importance for the acceptance. Algorithmic solutions must become a natural part of the team and a useful tools of the people, that actually helps them to perform their work. One important factor will be the freedom of decision. Even if an AI algorithm has found an optimum solution, it could be necessary for the overall acceptance, to hide parts of this solution and leave a remaining step of the decision process with the human workers. The AI can hereby supply additional information, to justify the decision later on. It can even use persuasive techniques so that the operator reaches the optimum result on his own.

Table 1. Potential fears of workers as challenge.

Fear of	Possible thoughts of worker	Countermeasure
being outperformed	• AI provides better production results • AI apparently has better understanding of production	• Make AI a tool • Avoid competition between AI and workers • Keep workers in the decision process
changing work focus	• With better AI performance, the focus of work will shift to other aspects • New work profile requires additional training	• Guide workers to new profiles • Make a transparent plan for the staff • Develop onboarding strategy for AI solutions • Provide training possibilities
becoming obsolete and losing the job	• AI was introduced to safe costs • Staff is mayor cost factor	• AI must be overseen by experienced operators • Define new roles

2.5 Social Challenges Caused by Expert Acceptance

A very different type of fear that may arise is related to expert knowledge. Often, AI solutions are considered pure black-boxes. While this is not explicitly the case for decision trees, forest approaches or Bayesian networks, for most neural network the decision path of the algorithm is hard to comprehend. Nevertheless, the automatic finding of results by a computer will always have to face scepticism, criticism and denial, if it is not possible to understand why or how a certain result was obtained. Clarity, transparency and plausibility are important factors for the acceptance of AI in the expert domain.

This problem was recognized already some time ago and the term explainable AI was coined. The first part of explainable AI considers easy reproducibility of results. Especially for neural networks this is an essential constraint, as the training of a network does sometimes involve randomisation of weight matrices and biases. Secondly, a translation layer is needed, to transform the operation and results of the AI method into domain specific language. Here, semantic descriptions play a crucial role. With a semantic process description, it is possible to interpret the results of algorithms in a human readable form. Details on this parallel semantic layer are shown in Fig. 4.

The semantic approach is also relevant for another, currently strongly rising technology: physics-informed AI. Although the original intention of machine learning aimed at automatisation of the knowledge extraction, adding expert know-how to the training and to the method topology, helps to accelerate the learning speeds and even reduces the amount of necessary input data for training. The reason for this can be seen by the following example: Assume a neural network is used as regression model, expected to match a cooling curve. If the network is aware about the physical fact, that cooling often exhibits an exponential form, the training can exploit this knowledge and focus on potential differences from this rule.

A second example comes from the steel process chain, namely at the continuous casting process. If the mould level oscillates, this can point towards defects in the casted slabs. If any machine learning technique is trained on the raw data from the mould level, it is not very likely that a meaningful relationship between defects and mould level might be found. If instead the mould level time series are Fourier transformed or wavelet transformed, prior to being inserted into the ML method, the prediction is very reliable.

This shows, that the use of curated and assorted input data, as well, as the provision of mathematical equations to the methods can help to improve the results of an AI system. In [14] such physics-informed strategies were also shown for the use case of anomaly detection in process variables.

3 Frontiers for AI

In Fig. 5, some potential frontiers are listed and the current state-of-the-art is compared with an envisioned future situation, that allows are more substantial exploitation of AI technology.

3.1 Models

Already today, standalone models of processes play an important role for the overall forecasting. AI, especially physics-informed AI as depicted in the Sect. 2.5, will help to improve these models and to incorporate stochasticity of the processes. This is considered a current frontier in the implementation of AI methods in process industry. It is important for future works to include probabilistic effects in the model generation as good as possible, as these are well preserved in the highly resolved time series data. The exploitation of these probabilistic perspective - the analysis of the process stochasticity - has up to now not fully been embraced by the industry, although many root-causes for defects and inferior production quality can be expected to associated with this aspect.

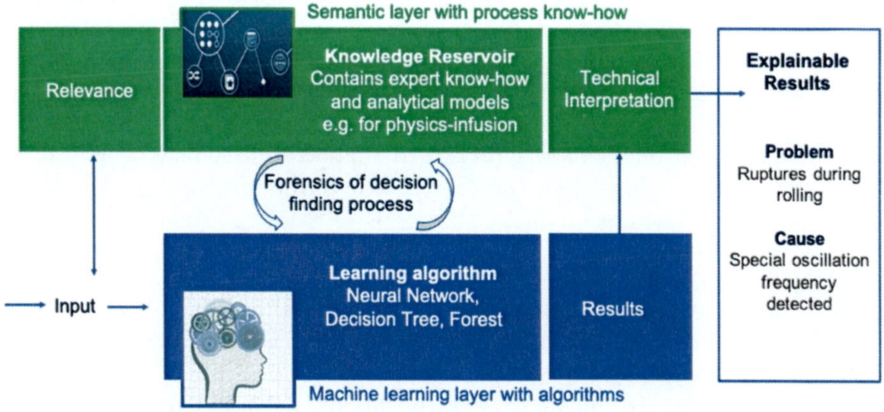

Fig. 4. One step closer to real AI: example to realize explainable AI solutions by utilizing a semantic layer with expert knowledge.

3.2 Forecasting of Production Chain

Novel simulation and forecasting techniques for the whole production chain will consider series and combinations of the above models. Technically, they will use massively parallelized evaluation, allowing distributed calculation and quick forecast times. The quality of those production forecasts must be improved, e.g. by research activities. Chains of uncertain AI models must be constructed, propagating the result and the assumed errors of the result through the process chain.

Additionally, the probability distributions of the processes will depend on the path through the production corridor. Favourable and unfavourable segments of these distributions will help to avoid risk affine production situations.

3.3 Optimisation

One of the most important frontiers currently faced by process industry is optimisation. Since decades, optimisation techniques are used for forward planning, efficient scheduling and process refinement. But most of these solutions are rather static. Dynamic replanning, as it is necessary after unforeseen perturbations or downtimes, or rapid reaction strategies that are calculated online and in real-time, require well-suited process models to quickly forecast multiple potential future paths. AI provides such models. Moreover, AI can help to derive a plan that might not be statically optimised, but optimised for dynamic replanning.

3.4 Semantics

Semantic technology has already proven highly successful in many examples, for instance Hunt et al. show in [15] a system for paraphrase identification for plagiarism detection, exploiting knowledge extraction techniques. Sun and Liu present ways to infer the relation dynamics in social networks based on semantics in [16], which indicates how the technology is already advanced. In [17], Zillner et al. developed an ontological approach to support autonomous agents with contextual understanding about processes, component relationships and data sources, specialised for steel industry.

Based on these prior works one can deduce a straightforward procedure to realise such a semantic modelling:

- Step 1. Establish a controlled vocabularies of the processes following the Who, What, Why, When, Where and How approach.
- Step 2. Construct taxonomies based on the initial vocabularies and define relevant objects, classes and categories.
- Step 3. Build an ontology that interrelates the objects and defines relationships among the vocabulary dimensions.

Up to now, semantic modelling is still underdeveloped in process industries, although it is one of the most important construction sites: it provides contextual knowledge to computers, which can use this knowledge to cast educated decisions. Consequently, the semantic frontier is crucial to achieve true artificial intelligence.

To underline this fact, please refer again to Fig. 4. The sematic layer depicted in this figure was already discussed in Sect. 2.5. It is not only capable to make AI results interpretable, but more essentially it can be used to supply important knowledge to the AI for improved training and evaluation. If the AI has access to this knowledge, the amount of required training data reduces and the overall quality of AI model predictions improves, as e.g. shown in the physics-informed autoencoder techniques developed in [18].

4 Summary

Broad introduction of AI technology faces several challenges. The aim of this paper was, to collect and present these challenges. AI has many benefits, yet, there are multiple constraints, barriers and preconditions that must be overcome. We divided the challenges into four types, a) infrastructural, b) data related, c) security and d) social. Each type requires individual solutions, all of which are part of roadmap for comprehensive AI roll out in industrial context.

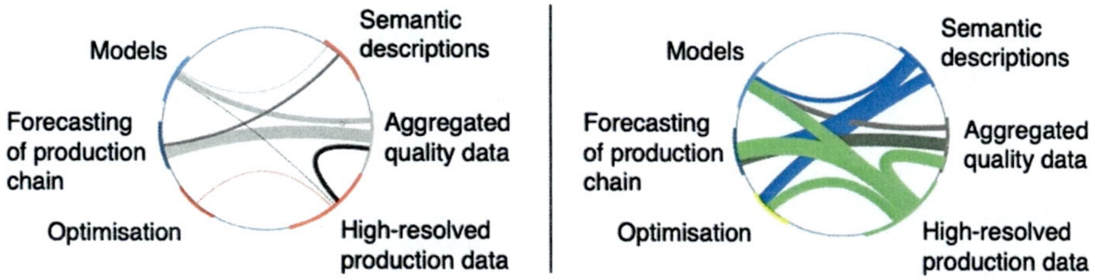

Fig. 5. Left: Situation as it is presumed to be. Right: Vision.

Technological frontiers are i) the exploitation of high resolution time-series data to improve modelling, ii) a smart combination of these models to achieve full process chain simulation and forecast capability, iii) a rigorous future utilisation of optimisation techniques that use these overall forecasts for finding the optimal solutions and iv) perhaps most importantly, the need to fundamentally embrace semantic modelling for introducing contextual knowledge to the computers, which can then extent current machine learning concepts towards real artificial intelligence.

References

1. Marz, N., Warren, J.: Big Data: Principles and Best Practises of Scalable Realtime Data systems. Manning Publications Co., Shelter Island (2015)
2. Chen, C.L.P., Zhang, C.-Y.: Data-intensive applications, challenges, techniques and technologies: a survey on big data. Inf. Sci. **275**, 314–347 (2014)
3. Court, D.: Getting impact from big data, Technical report McKinsey & Co (2016)
4. Aggarwal, S., Manuel, N.: Big data analytics should by driven by business needs, not technology, Technical report McKinsey & Co, June 2016
5. Geissbauer, R., Schrauf, S., Koch, V., Kuge, S.: Industry 4.0 - opportunities and challenges of the industrial internet, Technical report PricewaterhouseCoopers (PwC) (2014)
6. Arnu, D., Yaqub, E., Mocci, C., Colla, V., Neuer, M., Fricout, G., Renard, X., Mozzati, C., Gallinari, P.: A reference architecture for quality improvement in steel production. In: Proceedings 1st International Data Science Conference (iDSC) Data Science - Analytics and Applications, pp. 85–90 (2017)

7. Gzielo, A., Koszkul, J., Kwiatkowski, D., Pietrzak, M., Swierczynski, H.: Quality control system for the process of continuous casting of steel. J. Achieve. Mater. Manuf. Eng. **17**, 333–336 (2006)
8. Konrad, B., Lieber, D., Deuse, J.: Striving for zero defect production: intelligent manufacturing control through data mining in continuous rolling mill processes. In: Windt, K. (eds.) Robust Manufacturing Control. Springer, Cham (2013)
9. Kejela, G., Esteves, R.M., Rong, C.: Predictive analytics of sensor data using distributed machine learning techniques. In: IEEE Proceedings 6th International Conference Cloud Computing Technology and Science (CloudCom), pp. 626–631 (2014)
10. Berral-Garcia, J.L.: A quick view on current techniques and machine learning algorithms for big data analytics. In: IEEE Proceedings International Conference Transparent Optical Networks (ICTON), pp. 1–4 (2016)
11. Chen, X.W., Lin, X.: Big data deep learning: challenges and perspectives. IEEE Access **2**, 514–525 (2014)
12. Gao, Y., Shi, B., Dong, B., Chen, Y., Mi, L., Huang, Z., Shi, Y.: RVAE-ABFA: robust anomaly detection for highDimensional data using variational autoencoder. In: 2020 IEEE 44th Annual Computers, Software, and Applications Conference (COMPSAC), pp. 334–339 (2020)
13. Xu, X., Li, J., Yang, Y., Shen, F.: Towards effective intrusion detection using log-cosh conditional variational autoencoder. IEEE Internet Things J., p. 1 (2020)
14. Neuer: Automation security in the era of industry 4.0. Future Steel Forum Brochure, pp. 49–51, October 2019
15. Hunt, E., Janamsetty, R., Kinares, C., Koh, C., Sanchez, A., Zhan, F., Ozdemir, M., Waseem, S., Yolcu, O., Dahal, B., Zhan, J., Gewali, L., Oh, P.: Machine learning models for paraphrase identification and its applications on plagiarism detection. In: 2019 IEEE International Conference on Big Knowledge (ICBK), pp. 97–104 (2019)
16. Sun, W., Liu, X.F.: Inferring relationship semantics in social networks with dual-view features semi-supervised learning. In: 2019 IEEE International Symposium on Circuits and Systems (ISCAS), pp. 1–5 (2019)
17. Zillner, S., Ebel, A., Schneider, M.: Towards intelligent manufacturing, semantic modelling for the steel industry. IFAC-PapersOnLine **49**(20), 220–225 (2016). 17th IFAC Symposium on Control, p. 2016. Optimization and Automation in Mining, Mineral and Metal Processing MMM
18. Neuer: Quantifying uncertainty in physics-informed variational autoencoders for anomaly detection. In: ESTEP AI&ML Workshop (2020)

Data Pre-processing for Efficient Design of Machine Learning-Based Models to be Applied in the Steel Sector

Silvia Cateni(✉), Valentina Colla, Antonella Vignali, and Marco Vannucci

TeCIP Institute, ICT-COISP Center,
Scuola Superiore Sant'Anna, 56124 Pisa, Italy
{silvia.cateni,valentina.colla,antonella.vignali,
marco.vannucci}@santannapisa.it

Abstract. Data pre-processing plays a significant role in data mining in ensuring good quality data especially when dealing with industrial datasets. Dimensionality reduction and outliers detection are significant tasks in machine learning and data mining applications. Many industrial real-world datasets are affected by several errors and present a huge number of process variables. This circumstance can cause problems when the data are exploited to develop efficient machine learning-based models. In this paper some automatic pre-processing procedures are proposed considering both detection of anomalous data and informative variables selection to be used before models training in order to improve their performance. The paper proposes three industrial case studies coming from the context of the steelmaking industry and all the obtained results prove the efficiency of a correct and appropriate pre-processing stage.

Keywords: Data mining · Data Pre-processing · Outlier detection · Variable selection · Industrial dataset

1 Introduction

Feature space reduction and anomalous data detection are significant pre-processing phases in machine learning and data mining applications. Many real-world datasets are affected by errors, due to different reasons, and often contain redundant or not significant variables. These facts have negative effects in modelling applications, when Machine Learning (ML) techniques, such as, for instance, Neural Networks (NN), are applied to reproduce an unfamiliar relationship or phenomenon. Moreover, the data quality and a careful selection of the input variables become essential for the development of a reliable model. Anomalous patterns, also called *outliers* [1], are patterns that deviate from the rest of available data. They can be due to different reasons, such as measurements errors or variability of the phenomenon under consideration. The outlier detection is a central pre-processing step in various applications, such as financial analysis, fraud detection, network robustness analysis, network intrusion detection and marketing [2–5]. On the other hand, dimensionality reduction involves the selection of

V. Colla and C. Pietrosanti (Eds.): ESTEP 2020, AISC 1338, pp. 13–27, 2021.
https://doi.org/10.1007/978-3-030-69367-1_2

the most significant variables, which represent a subset of informative and uncorrelated input variables that help the improvement of both model accuracy and knowledge of the phenomenon under observation [6]. Variable selection is particularly significant when dealing with industrial datasets, especially when the number of input variables is considerable with respect to the number of samples. For instance, in the industrial context often a large number of process variables are measured by sensors, which are distributed along the production chain [7]. The issue of variable selection is intensely investigated in literature for conventional purposes such as prediction [8–10], classification [11–16] or clustering [17].

The present paper proposes three industrial case studies coming from steelmaking industrial context and all the obtained results demonstrate the effectiveness of an accurate and suitable pre-processing phase. The first industrial case study refers to the classification of a particular type of surface defects on flat steel products based on steel chemistry, some parameters of the steelmaking and continuous casting stages and proves the importance of variable selection for achieving acceptable model accuracy as well as understanding of the phenomenon under consideration. The second industrial problem refers to the study of the occurrence of ripple defects in the Hot Deep Galvanising (HDG) process for flat steel products and is focused on the study on thick coils having low zinc coating. Even if skilled personnel can control ripples defects occurrence through particular operations, such as wiping nitrogen as a substitute of air in air blades, the real effects of each process parameter variation are not identified. A deep investigation of the occurrences of these defects can contribute to the improvement of the quality of coils, by reducing reworked or scrapped material and by decreasing costs due to an excessive use of nitrogen. In this case, an appropriate pre-processing procedure has been performed and then the analysis was focused on the possible causes of ripples occurrences, focusing on the identification of process variables which mainly affect the presence or absence of ripples. In the third considered industrial case study, the main objective is the prediction of an important mechanical property of flat steel products undergoing the HDG process by exploiting the available process parameters. The target is measured through a popular online measuring system for material properties, i.e. the Impulse Magnetic Process Online Controller (IMPOC). Also in this case, the paper shows the importance of an effective and accurate pre-processing stage for the design of Machine Learning-based forecasting models.

The paper is organised as follows: in Sect. 2 a review of traditional data pre-processing and dimensionality reduction techniques is provided. Section 3 describes the proposed approaches exploited in the three different industrial case studies. Section 4 presents the experimental results, while Sect. 5 provides some concluding remarks.

2 Typical Pre-processing Stages

One of the most important themes connected with all sorts of data analysis is data quality. The ongoing digitalization process in all the industrial sectors, including the steel sector [18], provides huge volumes of data which covey a relevant portion of the companies' know-how. However, real data also frequently include noise [19], uncertainty, errors, redundancies or irrelevant information, while the accuracy of the developed ML-based models which exploit those data for their training highly depends on data quality. This is the reason of why an accurate data cleaning could lead to an improvement of the efficiency of the exploited model. Data cleaning concerns both the elimination of outliers and the reduction of the variable space in order to develop models using only the "informative" variables.

2.1 Outlier Detection

Traditional outlier detection techniques can be classified into four main categories: distribution-based [20], distance-based [21, 22], density-based [23] and clustering-based [24]. Distribution-based methods are typically exploited in statistical applications [25]. According to this approach, an outlier is defined as a sample that does not accurately fit a standard distribution. The most common distribution-based approach is the Grubbs test, which uses as distribution the Gaussian function. The distance-based approach, introduced by Knorr and Ng in 1998 [26], provides the following definition of outlier: *"An object x in a dataset T is a DB(p,D)-outlier if at least fraction p of the objects in T lie at a distance greater than D from x"*. This approach does not require any a-priori assumption on the data but it is necessary to define a distance definition. An effective distance-based approach exploits the Mahalanobis distance. Clustering-based approaches define an outlier as a datum, which is not included in any cluster after a suitable clustering procedure. A common clustering method is the Fuzzy C-Means (FCM) proposed by Dunn and improved by Bedzek [27]. Finally, density-based approaches [23] attribute to each data sample a degree of "outlierness", the so called Local Outlier Factor (LOF). The LOF specifies how a data point is isolated with respect to its neighborhood: if the LOF is high then the sample is an outlier. Recently, fuzzy logic has been exploited in some outlier detection methods applied in many real industrial problems [28]. For instance, in [29, 30] a combination between an outlier detection method and a clustering algorithm has been proposed. The outlier detection method is applied to establish if a sample can be considered as an outlier and the clustering algorithm allocates samples to clusters. The two results are subsequently combined in order to deliver original measure of outlierness. In [31] a fuzzy inference system is used to combine several traditional outlier detection techniques. This method is able to exploit the advantages of the traditional methods avoiding their drawbacks [32].

2.2 Redundant Variables Elimination

The selection of the most informative input variables generates many benefits, such as lower complexity of the model and of the learning algorithm, savings of computational time and increased performance. Moreover, selecting the informative variables can lead to acquire a deeper knowledge of the investigated problem by detecting the factors, which mostly affect it. A first stage consists in removing high correlated variables in order to exclude redundancy in the dataset, while subsequently a variable selection procedure must be applied in order to select the variables, which mostly affect the considered target. Redundant variables can be recognized by exploiting the dominating set algorithm [33]. Such algorithm, which derives from graph theory, builds a sort of graph *G*, where the vertexes correspond to the variables, and establishes a link among two vertexes if the corresponding variable show a linear correlation coefficient greater than 0.95 and a p-value lower than 0.05 (which means that the computed correlation among the two considered variables is significant) [34]. Then, the algorithm selects the minimum dominating set of the graph as the subgraph *G'* where every vertex of *G* not included in *G'* is adjacent to at least one member of *G'*. Redundant variables are those ones, whose associated vertexes do not belong to *G'* [35]. Once redundant variables have been eliminated, the procedure for selecting the most significant variables with respect to the target to be predicted follows [36, 37].

2.3 Variable Selection

Variables selection methods can be divided in 3 main classes: filter, wrapper and embedded. Filter approaches [38–42] use statistical tests to achieve the selection by providing a pertinence score for each variable and identifying the variables which obtain a high score. Filter methods are independent on the learning algorithm developed and their computational burden is usually low. A common filter method is the *correlation-based* approach, which evaluates the correlation coefficient between each variable and the target to be predicted; variables are subsequently ranked and a subset is extracted containing the variables with the highest correlation coefficient, also introducing an appropriate threshold. This method is fast but not effective when the variables are not linearly correlated to the target. Other commonly exploited filter approaches are the *Fisher criterion* and the *T-test*, which are applied to establish the significance of each variable in the case of binary classification, considering the mean value and the standard deviation of the samples belonging to the two classes. Another filter approach exploits the concept of Mutual Information (MI). MI quantifies the amount of information conveyed by a variable or a set of variables for the prediction of a given target. MI does not depend on the model linking the input variables to the target and is nonlinear [43].

Wrapper approaches exploit the performance of the learning machine in order to select a subset of variables estimating their predictive power considering the learning algorithm as a black box. These approaches are universal, as they can be exploited in combination with different learning algorithms [44, 45]. An obvious wrapper approach is the exhaustive search, also called *brute force* approach, as it evaluates all the variables combinations. When the number n of input variables is significant, the exhaustive search is impractical, as it requires the execution of 2^n model training procedures. A common wrapper approach is the so-called *Greedy Search* strategy, which progressively generates the variables subset by adding or removing single variables from an initial set. Greedy search can be applied in two directions: *Sequential Forward Selection (SFS)* and *Sequential Backward Selection (SBS)*. SFS starts from an empty set of variables and iteratively incorporates other input variables until a fixed stopping criterion is achieved. SBS works in the opposite direction: the procedure starts including all features and progressively eliminates the least relevant ones. A variable is considered relevant if the performance of the learning machine tends to get worse when it is removed. Greedy search strategies require at maximum $n(n + 1)/2$ training steps and, as SFS begins with a minor variables set, it is less expensive than SBS, if it ends early. Another efficient wrapper method is based on Genetic Algorithms (GA) [8, 14, 15]. The chromosomes represent the variables subsets and are binary vectors, whose size equals the number of input variables being each gene related to a variable, so that it is unitary when the associated variable belongs to the considered subset. The GA population is randomly initialized and genetic operators, such as mutation and cross-over, are calculated at each generation. The mutation operator switches a randomly selected gene, while the crossover operator generates each gene of the son chromosome by mixing the genes of the two parent chromosomes. Before applying the variable selection process, the dataset is divided in two subsets: 75% of the data is associated to the training set, while the remaining 25% of data is exploited as validation set. The fitness function to be minimized is the mean relative prediction error of the model and is estimated on the validation set. The GA ends when a fixed maximum number of iterations has been achieved or a plateau of the fitness function is reached. The winner chromosome identifies the subset of input variables associated to the best value of the fitness as well as the model exploiting that subset.

Embedded methods perform the variable selection as part of the learning phase and are logically specific of a precise learning algorithm [46–48]. Typical examples of embedded methods are classification trees, random forests. The main advantage of embedded methods lies in their inclusion in the learning algorithm, which considers the relevance of the variables.

To sum up, filter methods are suitable and effective approaches to treat with very high-dimensional datasets; they are computationally simple and fast as they are independent on the algorithm used. Wrapper methods consider the classifier as a black box using its performance for the selection of variables. On the other hand, wrappers are computationally heavier than filters, mainly if the datasets are vast or if the classifier has a high computational cost. Finally, embedded methods have a lower computational cost than the wrappers but they are too specific for the exploited classifier.

In order to exploit most of the advantages of the various methods while avoiding their drawbacks, the so-called hybrid models have been proposed in recent years. An example of hybrid model is proposed in [49, 50]. This procedure starts with a whole dataset and reduce it, exploiting a combination of popular filter methods, in order to obtain a minimal dataset including only informative variables. Subsequently, the exhaustive search technique is performed in a reasonable time in order to reach a sub-optimal subset of features. The indication to decrease the feature space and then to apply the exhaustive search was proposed by Lee in [51]. Each filter method outputs a score of importance for each variable. Scores are firstly normalized in order to achieve values in the range [0, 1] and then their average value is computed. The final selection is performed through a threshold decider fed with the average score.

3 Steel Industrial Applications

3.1 First Case Study: Inclusion Classification

The first case study regards the classification of particular surface defects named *inclusions*, whose presence affects the quality of the products: the higher the number of inclusions, the lower the quality. An inclusion is a very small particle of external material inside the steel matrix, whose occurrence causes surface defects when the material is rolled [52, 53]. Considering their origin, inclusions can be classified as exogenous or endogenous, while observing their size, they can be classified as micro or macro inclusions. Micro/macro inclusions can be both exogenous and endogenous. Micro inclusions show a size lower than 5 μm and are mainly caused by a reaction between oxygen and other additions in liquid steel. Macro inclusions are larger than 25 μm and are formed generally in tundish due to chemical reactions, re-oxydation, slag entrapment or refractory erosion.

The objective of this study is to classify surface defects as inclusions or not, considering parameters related to the steel chemistry or to the steelmaking and continuous casting processes. To this aim, a hybrid variable selection approach [49] is applied: a preliminary combination of filter approaches is performed to reduce the number of variables and then the wrapper exhaustive search technique is implemented in a reasonable time to achieve a sub-optimal subset of features considering the accuracy of the classifier [54]. The initial available dataset included 18 variables listed in Table 1.

After the reduction of the dataset through filters combination, data are divided in training (75% of all samples) and validation set (the remaining 25%). Five common and widely used in literature classifiers have been applied: a Bayesian Classifier (BC), a Decision Tree (DT), a Support Vector Machine (SVM), a Multi-Layer Perception (MLP) and a Self-Organising Map (SOM). The importance of the preliminary variable selection step can be assessed by using the validation dataset to estimate the performance of all classifiers. The 10-cross validation technique is exploited in evaluation phase and the mean performance is taken into account. The average accuracy, also known as Balanced Classification Rate (BCR) [54, 55], shown in Eq. 1, is evaluated as measure of accuracy, since it is appropriate in case of imbalanced classes.

$$BCR = \frac{1}{2}(sensitivity + specificity) = \frac{1}{2}\left(\frac{TP}{TP+FN} + \frac{TN}{TN+FP}\right) \quad (1)$$

where true positive (TP) and true negative (TN) are, respectively, positive and negative samples which are correctly classified, while, false negative (FN) and false positive (FP) represent, respectively, misclassified negative and positive samples. This accuracy index is the arithmetic mean of two commonly used metrics: *sensitivity*, which measures the proportion of positives that are correctly classified, and *specificity,* which measures the proportion of negatives that are correctly identified.

Table 1. Variables included in the initial datasets

Variables	
1.	Reheating Temperature
2.	Blowed Oxygen (nm^3)
3.	Aluminum content (Kg)
4.	Duration of the reheating process (min)
5.	Oxygen in ppm measured at the treatment
6.	O_2 in the converter (mn^3)
7.	Silicon variation (difference between Si during continuous cast and Si in ladle)
8.	Nitrogen variation
9.	Phosphorous variation
10.	Mean casting speed line 1 (cm/min)
11.	Mean casting speed line 2 (cm/min)
12.	Total number of slabs
13.	Total weight of slabs
14.	Tapping temperature
15.	Ladle temperature
16.	Treatment temperature
17.	Casting temperature
18.	Tundish temperature

3.2 Second Case Study: Ripple Defect Detection

The second case study is devoted to the detection of ripple defects during the Hot-Dip Galvanizing (HDG), concentrating the study on thick coils with low zinc coating [56, 57]. HDG is one of the most exploited processes for coating steel with a layer of zinc. This practice comprises the passage of the steel strip through molten zinc at a temperature of 860 F in order to allow formation of zinc carbonate ($ZnCO_3$). Zinc carbonate is a strong material that offers protection against corrosion over the steel surface.

Ripples appear as transversal lines and they are identified by an Automatic Surface Inspection System (ASIS). The occurrence and severity of this kind of defect can be affected by particular process variables. Even though skilled personnel can manage ripple defects with specific operations, such as wiping nitrogen as a substitute of air in air blades, the actual effects of each process parameter variation is not precisely known. Therefore, the study of these phenomena can improve the quality of coils, by reducing reworked or scrapped material and decreasing costs related, e.g. to an excessive use of nitrogen. Data concerning 356 coils are available, while the 20 available process variables belong to 4 macro categories: air blades parameters (use of nitrogen, pressure, distance and angle); temperatures (zones before and after the zinc bath, top-roll, water bath); line speed and fan coolers parameters along the cooling tower. Quality and process data are collected along the entire production chain: some process variables are sampled every 10 meters of coil, while the other ones are sampled every 1 m. Each sample represents a longitudinal position on the coil, while defects, detected by ASIS, are represented by a two dimensional spatial position on the coil and by the considered area. Therefore, the coil surface is divided into tiles corresponding to portions of the coil. 177 coils underwent the nitrogen blowing, while 179 coils underwent air blowing. Therefore, two different datasets for air and nitrogen blowing are created.

After the datasets design, an accurate pre-processing stage is implemented. Firstly, a plausibility investigation is executed in order to work with variables having acceptable values. Subsequently, the automatic fuzzy-based outlier detection stage is performed. Afterwards, the analysis focuses on the identification of potential reasons of ripples occurrences. In order to label coils slices, a threshold is fixed to the 95th quantile of the distribution of defect occurrence. Slices with a number of defects overcoming that threshold are classified as defective (class 1), the other ones are classified as not defective (class 0). For both datasets a training and a validation set (formed by the 75% and the 25% of the whole data set, respectively) are created by randomly shuffling the original dataset but maintaining the original proportion among the two classes. For each case study a Decision Tree (DT) is implemented and then valued on the respective validation set. Classification performance are evaluated through the BCR. DTs are exploited for two reasons: they can be interpreted by non-expert persons, as they appear as a simple chain of IF-THEN-ELSE rules, by thus representing a support to decisions. Moreover, they are a typical embedded technique for variable selection. The algorithm to calculate feature importance is integrated as part of the learning algorithm; it selects a variable in each recursive step of the tree growth process and divides the sample set into smaller subsets. Feature importance is evaluated as the decrease in node impurity weighted by the probability of reaching that node. The node probability can be computed by the number of samples that reach the node, divided by the total number of samples. The higher this value, the more important the variable.

3.3 Third Case Study: Mechanical Property Prediction

This case study concerns cold rolled strips, rolled up as coils, which comes from the pickling line and the Cold Rolling Mill (CRM) and are processed through a HDG line to cover them with a zinc coating. Here an accurate pre-processing stage was implemented before the development of a model forecasting a significant mechanical property, which is detected by the IMPOC system [58]. Samples of available dataset have been stored during the HDG process and they are related to longitudinal positions on the coil having a spatial interval of 10 m. The investigation is focused on three steel grades, which are indicated here as SG_1, SG_2 and SG_3 for confidentiality constraints. The procedure starts from a real raw dataset and provides a prediction of a mechanical property of the final steel product after an accurate pre-processing phase, which leads to the design of computationally efficient and accurate prediction models.

A first variable selection phase is based on the process knowledge of the experts and plant managers. Then a plausibility study is performed in order to create reliable datasets having process parameters within tolerable value ranges. All samples or coil positions showing at least one value missing or not plausible are deleted and if more than 30% of the measurements of a variable are not plausible, that variable is removed. Moreover, only reliable coils are kept in the cleansed data set, namely the coils having plausible measurements for at least 80% of coil positions. Subsequently, a fuzzy-based outlier detection is applied in order to detect all anomalous data. In addition, another check is executed concerning chemical uniformity of coils. Only coils resulting uniform from a chemical point of view are selected using the following definition: a coil is defined uniform if its chemical analysis can be considered almost constant (by means of mean and standard deviation of chemical elements) towards the chemical analysis of all coils belonging to the same steel code.

Once reliable coils are identified, the most informative process variables are selected. Firstly, a redundancy analysis is performed through the previously mentioned dominating set algorithm. Finally, a GA-based variable selection procedure is performed. Different models are generated according to steel code and thickness, which are elements, which mainly affect the predicted mechanical properties. Therefore, coils are divided into several classes considering the steel code and each class has been further divided in 10 sub-classes associated to different thickness ranges of width 0.2 mm. For each sub-class three prediction models have been developed: a polynomial approximation and two 3-layered feedforward NN. The number of neurons in the NNs hidden layers is empirically selected based on several tests.

4 Results

4.1 Results on First Case Study

The hybrid variables selection selected only 4 variables in the final subset: blowed Oxygen, Nitrogen content in the liquid steel, first speed of the casting and ladle temperature. Figure 1 shows the results in terms of BCR in two cases: by exploiting all the available variables and then after the application of the variable selection procedure.

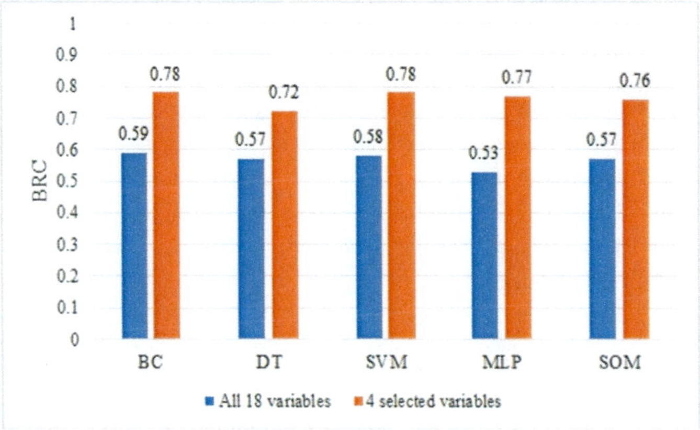

Fig. 1. Results of the first case study

The performance of all the classifiers is remarkably improved by the hybrid variable selection procedure. The application of the wrapper variable selection procedure provides not only a well performing classifier but also improves the knowledge of the considered phenomenon.

4.2 Results on Second Case Study

After the above-described pre-processing phase, the two datasets, obtained with air or nitrogen blowing, are separately fed to a DT-based classifier. The average accuracy of the binary classification is estimated in terms BCR and then the mean accuracy on 10 iterations taking different training sets is computed. The results in the two analyzed cases (air and nitrogen blowing) are displayed in Table 2.

Table 2. Obtained results for the second case study.

Blowing type	Overall accuracy %	Acc. class 0%	Acc. class 1%
Air	99.34	99.78	98.90
Nitrogen	97.65	98.92	96.37

Satisfactory classification performances in terms of average accuracy are achieved. Moreover, thanks to definite rules inferred by DTs, it is possible to manage the affecting variables in order to obtain the required quality level. Finally, the most informative variables are identified. Tables 3 and 4 report the mean importance degree of the significant variables in a decreasing order in the two analyzed cases: air blowing and nitrogen blowing. The listed variables are in line with the experts' expectations.

Table 3. Mean variable importance degree for air blowing.

Process variables	Mean importance degree
Distance air blade	1
Tunnel zone temperature	0.69
Speed process section	0.29
Pressure air-blade	0.25
Fans reference speed	0.15
Top-roll zone temperature	0.08
Water bath cooling temper.	0.07
Strip temperature	0.02

Except for the Water bath cooling temperature (which is a consequence of other parameters such as residual strip heat, speed and thickness) all other selected variables can be adjusted in line to tune the process under "ripple safe" conditions.

Table 4. Mean variable importance degree for Nitrogen blowing.

Process variables	Mean importance degree
Water bath cooling temper.	1
Distance air blade	0.45
Hot bridle zone temper.	0.39
Speed process section	0.14
Pressure air-blade	0.14
Fans reference speed	0.10
Height air knife	0.05
Top-roll zone temperature	0.02

4.3 Results on Third Case Study

Performance results have been evaluated taking as target the Ultimate Tensile Strength (UTS), expressed in Mega Pascal (MPa). Through the preliminary knowledge-based variable selection stage, 37 input variables have been selected. By executing the plausibility analysis and then the above described outlier detection approach no variable was excluded but about 30% of coils has been discarded from the dataset. The results of the preliminary uniformity analysis indicated that the 72.8%, 77% and 78.6% of the data related to the chemical content of all coils, respectively, for the three considered steel grades are uniform. Afterwards, starting from reliable data, the redundant variables are identified and eliminated. Finally, the previously mentioned GA-based variable selection procedure is implemented.

The obtained results are in line with the expectation of the technical experts of the company. The performances of two NNs, the first one trained using the raw data set and the other one trained with the cleansed dataset (referring to the same period) are compared in order to prove the efficiency of an accurate and appropriate data pre-processing stage. The results obtained by training the network with the raw and the cleansed datasets are reported in Tables 5 and 6 in terms of average error ε and standard deviation σ as well as average relative error ε_r and standard deviation σ_r.

Table 5. Results obtained by training the NN with the raw dataset.

Steel grade	ε	σ	ε_r	σ_r
SG_1	34.78	12.11	0.118	0.041
SG_2	63.75	52.18	0.163	0.134
SG_3	87.72	111.16	0.193	0.248

Table 6. Results obtained by training the NN with the cleansed dataset.

Steel grade	ε	σ	ε_r	σ_r
SG_1	5.51	4.81	0.019	0.016
SG_2	10.25	9.73	0.026	0.027
SG_3	23.38	16.74	0.056	0.044

This analysis proves the importance and the effectiveness of the data preprocessing stage in improving the accuracy of the NN-based models: the performance achieved when data pre-processing is implemented improves on average more than 20%. Moreover, the selection of the most informative variables improves the knowledge about the investigated process, as the process parameters mostly affecting UTS are identified.

5 Conclusions

The paper overviews the main pre-processing stages, which are necessary in order to successfully implement classical ML-based modelling and classification tasks. Moreover, it shows the importance of an adequate pre-processing stage through the discussion of three exemplar applications related to the steel sector. In particular, the first case study deals with the classification of particular surface defects of flat steel products on the basis of the steel chemistry and some process parameters. The second case study concerns the occurrence of ripple defects during HDG in the production of flat steel products. Here the pre-processing is preliminary to a DT-based classification, which provides clear rules that are useful to support decisions. The third case study concerns the design and development of NN-based models for the prediction of a relevant mechanical property of flat cold-rolled and hot dip galvanized products.

The results obtained in all the considered cases study highlight the advantages, which can be achieved through the application of an adequate data pre-processing in terms of both model performances and improved knowledge on the considered process or phenomenon. This conclusion can be extended to any ML-based approach, as data-driven approaches are also data-dependent, namely the goodness of the results strongly depends on the quality of the exploited data.

References

1. Hawkins, D.: Identification of Outliers. Chapman and Hall, London (1980)
2. Han, J., Kamber, M.: Data Mining Concepts and Techniques, Morgan Kauff. Publ., Burlington (2001)
3. Vannucci, M., Colla, V., Corbo, G., Fera, S.: Detection of rare events within industrial datasets by means of data resampling and specific algorithms. Int. J. Simul. Syst. Sci. Technol. **11**(3), 1–11 (2010)
4. Bakar, Z., Mohamed, R., Ahmad, A., Deis, M.: A comparative study for outlier detection techniques in data mining. In: Conference on Cybernetics and Intelligent Systems, Bankok, pp. 1–6 (2006)
5. Hodge, V.J.: A Survey of Outlier Detection Methodologies. Kluwer Academic Publishers, Dordrecht (2004)
6. De Lucena, A., Soares, A., Filho, A.: Multiobjective evolutionary algorithm for variable selection in calibration problems: a case study for protein concentration prediction. In: IEEE Congress on Evolutionary Computation, Cancun Mexico, 20–23 June 2013

7. Vannucci, M., Colla, V., Sgarbi, M., Toscanelli, O.: Thresholded neural networks for sensitive industrial classification tasks. Lecture Notes in Computer Sciences, LNCS, PART 1, vol. 5517, pp. 1320–1327 (2009)
8. Cateni, S., Colla, V., Vannucci, M.: General purpose input variables extraction: a genetic algorithm based procedure GIVE A GAP. In: 9th International Conference on Intelligent Systems Design and Applications (ISDA), pp. 1278–1283 (2009)
9. Guillon, A., Sovilj, D., Lendasse, A.: Minimising the delta test for variable selection in regression problems. Int. J. High Perform. Syst. Architect. **1**(4), 269–281 (2008)
10. Sofge, DA., Elliot, D.L.: Improved neural modeling of real-world systems using genetic algorithm based variable selection. In: Proceedings Conference on Neural Networks & Brain (1998)
11. Sofge, D.A.: Using GA based variable selection to improve neural network models for real-world systems. In: Proceedings of the International Conference on Machine Learning & Applications (2002)
12. Xu, L., Zhang, W.G.: Comparison of different methods for variable selection. Anal. Chim. Acta **446**, 477–483 (2001)
13. Lin, J., Ke, H., Chien, B., Yang, W.: Classifier design with feature selection and feature extraction using layered genetic programming. Exp. Syst. Appl. **34**, 1384–1393 (2008)
14. Cateni, S., Colla, V., Vannucci, M.: A genetic algorithms-based approach for selecting the most relevant input variables in classification tasks. In: Proceedings UKSim 4th European Modelling Symposium on Computer Modelling and Simulation, EMS 2010, pp. 63–67, (2010)
15. Cateni, S., Colla, V., Vannucci, M.: A genetic algorithm-based approach for selecting input variables and setting relevant network parameters of a SOM-based classifier. Int. J. Simul. Syst. Sci. Technol. **12**(2), 30–37 (2011)
16. Colla, V., Bioli, G., Vannucci, M.: Model parameters optimisation for an industrial application: a comparison between traditional approaches and genetic algorithms. In: Proceedings - EMS 2008, European Modelling Symposium, 2nd UKSim European Symposium on Computer Modelling and Simulation, pp. 34–39 (2008)
17. Wang, S., Zhu, M.: Variable selection for model based high dimensional clustering and its application to microarray data. Biometrics **64**, 440–448 (2008)
18. Branca, T.A., Fornai, B., Colla, V., Murri, M.M., Streppa, E., Schröder, A.J.: The challenge of digitalization in the steel sector, Metals **10**(2), 288 (2020)
19. Sgarbi, M., Colla, V., Cateni, S., Higson, S.: Pre-processing of data coming from a laser-EMAT system for non-destructive testing of steel slabs. ISA Trans. **51**(1), 181–188 (2012)
20. Kwak, S.K., Kim, J.H.: Statistical data preparation: management of missing values and outliers. Korean J. Anesthesiol. **70**(4), 407 (2017)
21. Knorr, E.M., Ng, R.: Algorithms for mining distance based outliers in large datasets. In: Proceedings of VLDB, pp. 392–403 (2003)
22. Mahalanobis, P.C.: On the generalized distance in statistics. Proc. Nat. Inst. Sci. India **4**, 9–55 (1936)
23. Breunig, M., Kriegel, H., Ng, T., Sander, J.: LOF: identifying density based local outliers. SIGMOD Record **29**(2), 93–104 (2000)
24. Xue, Z., Shang, V., Feng, F.: Semi-supervised outlier detection based on fuzzy rough C-means clustering. Math. Comput. Simul. **80**, 1911–1921 (2010)
25. Barnett, V., Lewis, T.: Outliers in Statistical Data. 3rd edn., Wiley, Hoboken (1984)
26. Zhang, K., Hutter, M., Jin, H.: A new local distance-based outlier detection approach for scattered real-world data. In: Pacific-Asia Conference on Knowledge Discovery and Data Mining, pp. 813–822 (2009)

27. Cateni, S., Colla, V.: A hybrid variable selection approach for NN-based classification in industrial context. Smart Innov. Syst. Technol. **69**, 173–180 (2017)
28. Dettori, S., Iannino, V., Colla, V., Signorini, A.: An adaptive Fuzzy logic-based approach to PID control of steam turbines in solar applications. Appl. Energy **227**, 655–664 (2018)
29. Yousri, N., Ismal, M., Kamel, M.: Fuzzy outlier analysis. a combined clustering-outlier detection approach. In: Proceedings IEEE International Conference Systems, Man and Cybernetics SMC 2007, pp. 412–418 (2007)
30. Rousseeeuw, P., Leroy, A.: Robust Regression and Outlier Detection. Wiley, Hoboken (2003)
31. Cateni, S., Colla, V., Vannucci, M.: A fuzzy logic based method for outliers detection. In: Proceedings of the IASTED International Conference on Artificial Intelligence and Applications (AIA 2007), pp. 561–566 (2007)
32. Guyon, I., Elisseeff, A.: An introduction to variable and feature selection. Mach. Learn. **3**, 1157–1182 (2003)
33. Bondy, J.A., Murty, U.: Graph Theory, Springer, Cham (2008)
34. Yu, L., Liu, H.: Feature selection for high-dimensional data: a fast correlation based filter solution. In: Proceedings of the 20th International Conference on Machine Learning, ICML 1, pp. 856–863 (2003)
35. Cateni, S., Colla, V.: A hybrid variable selection approach for NN-based classification in industrial context. Smart Innov. Syst. Technol. **69**, 173–180 (2017)
36. May, R., Dandy, G., Maier, H.: Review of input variable selection methods for artificial neural networks. Artif. NN Methodolog. Adv. Biomed. Appl. **10**, 16004 (2011)
37. Liu, H., Motoba, H., Setiono, R., Zhao, Z.: Feature selection: an ever evolving frontier in data mining. In: Proceedings 4th Workshop on Feature Selection in Data Mining, pp. 4–13, (2010)
38. Mitchell, T., Toby, J., Beauchamp, J.: Bayesian variable selection in linear regression. J. Amer. Statist. Ass. **83**, 1023–1032 (1988)
39. Sun, Y., Robinson, M., Adams, R., Boekhorst, R., Rust, A.G., Davey, N.: Using feature selection filtering methods for binding site predictions. In: Proceedings 5th IEEE International Conference on Cognitive Informatics ICCI 2006, pp. 566–571 (2006)
40. He, X., Cai, D., Niyogi, P.: Laplacian score for feature selection. In: Advances in Neural Information Processing Systems, pp. 507–514 (2005)
41. Cateni, S., Colla, V., Vannucci, M.: A fuzzy system for combining filter features selection methods. Int. J. Fuzzy Syst. **19**(4), 1168–1180 (2016)
42. Zhang, S., Zhao, Z.: Feature selection filtering methods for emotion recognition in Chinese speech signal. In: 9th International Conference on Signal Processing, ICSP, pp. 1699–1702 (2008)
43. Cover., T., Thomas, A.: Elements of Information Theory. Telecommunications and Signal Processing. Wiley Series, Hoboken (1991)
44. Kohavi, R., John, G.H.: Wrappers for feature subset selection. Artif. Intell. **97**(1–2), 273–324 (1997)
45. Hsu, C.N. Huang H.J., Schuschel, D.: The ANNIGMA-Wrapper approach to fast feature selection for neural nets. IEEE Trans. Syst. Man Cybern. **32**(2), 207–212 (2002)
46. Bo, L., Wang, L., Jiao, L.: Multilayer perceptrons with embedded feature selection with application in cancer classification. Chin. J. Electron. **15**, 832–835 (2006)
47. Breiman, L.: Random forests. Mach. Learn. **45**, 5–32 (2001)
48. Breiman, L., Friedman, J.H., Olshen, R.A., Stone, C.J.: Classification and Regression Trees. Wadsworth and Brooks, Heidelberg (1984)

49. Cateni, S., Colla, V., Vannucci, M.: A hybrid feature selection method for classification purposes. In: Proceedings - UKSim-AMSS 8th European Modelling Symposium on Computer Modelling and Simulation, EMS 2014, pp. 39–44 (2014)
50. Cateni, S., Colla, V., Vannucci, M.: A fuzzy system for combining filter features selection methods. Int. J. Fuzzy Syst. **19**(4), 1168–1180 (2017)
51. Lee, K.: Combining multiple feature selection methods. Ph.D. thesis, The Mid-Atlantic Student Workshop on Programming Languages and Systems Pace University (2002)
52. Sheng, D.Y., Soder, M., Jönsson, P., Jonsson, L.: Modelling micro-inclusion growth and separation in gas-stirred ladles. Scandinavian J. Metallurgy **31**(2), 134–147 (2002)
53. Sheng, D.Y., Soder, M., Alexis, J.: Most relevant mechanism of inclusion growth in an induction-stirred ladle. Scandinavian J. Metallurgy **31**, 210–220 (2002)
54. Cateni, S., Colla, V.: The importance of variable selection for neural networks-based classification in an industrial context. Smart Innov. Syst. Technol. **54**, 363–370 (2016)
55. Sokolova, M., Lapalme, G.: A system analysis of performance measures for classification tasks. Inf. Process. Manage. **4**, 427–437 (2009)
56. Brandenburger, J., Colla, V., Nastasi, G., Ferro, F., Schirm, C., Melcher, J.: Big data solution for quality monitoring and improvement on flat steel production. IFAC-PapersOnLine **49** (20), 55–60 (2016)
57. Cateni, S., Colla, V., Vignali, A., Brandenburger, J.: Cause and effect analysis in a real industrial context: study of a particular application devoted to quality improvement. Smart Innov. Syst. Technol. **102**, 219–228 (2019)
58. Colla, V., Cateni, S., Maddaloni, A., Vignali, A.: A modular machine-learning-based approach to improve tensile properties uniformity along hot dip galvanized steel strips for automotive applications. Metals **10**(7), 923 (2020)

Quantifying Uncertainty in Physics-Informed Variational Autoencoders for Anomaly Detection

Marcus J. Neuer(✉)

Betriebsforschungsinstitut (BFI), Sohnstr. 65, 40237 Duesseldorf, Germany
marcus.neuer@bfi.de
http://www.bfi.de

Abstract. An example of anomaly detection for industrial production processes is shown, based on a variational autoencoder. Starting with a simple probability corridor, the concept of autoencoding is introduced and variational encoding is reviewed. With an input enrichment that utilizes difference optimization of a two class test set, the varational encoder is then extended to incorporate prior physical knowledge. Here, this knowledge is reflected by a weighted derivative series. Using a prototypical example from steel production, the standard probability approach, the autoencoder and the physics-informed, variational autoencoder are compared. The probability distribution of the latent space is visualized.

Keywords: Machine learning · Uncertainty quantification · Stochastic dynamics

1 Introduction

Nowadays, process data has become a primary resource for technical evaluations and business decisions. Redacted in the right way, it provides valuable insights to control, diagnose and finally optimise production processes. It is also fundamental for model development, because both white-box analytical models, derived from a first-principle relationship, and black-box machine learning models require those data for validation and training purposes.

But this data can not be seen as ground truth. It contains uncertainties, induced by different sources, most prominently by the precision and accuracy boundaries of sensors and measurement noise. All subsequent models built upon this data are, of course, influenced by such effects and therefore a comprehensive treatment of uncertainties is necessary.

Industrial production typically relies on more or less complex processes, with steel production as example, these include different heating, reheating and cooling stages, forming processes like hot and cold rolling, and finally refinement procedures like pickling and galvanisation. In all these steps, the product may show up deficiencies, which lead to costly refurbishment or complete recycling of the product later on. It is desirable for the companies to avoid such failures as good as possible.

V. Colla and C. Pietrosanti (Eds.): ESTEP 2020, AISC 1338, pp. 28–38, 2021.
https://doi.org/10.1007/978-3-030-69367-1_3

One class of methods that helps to detect failures is anomaly detection. As a matter of fact, this field is also benefiting from a proper understanding of the process variable uncertainties. If the normal process situation is understood well and, ideally, quantified in terms of allowed fluctuations, any divergence from this normality can be detected quickly.

Let us review some definitions from literature, that are relevant for our later discussion. Normally, we distinguish between so-called aleatoric and epistemic uncertainties. Aleatoric uncertainties refer to all those effects that introduce randomness in the measurement process itself and which can in principle not be overcome. In this sense, noise, random fluctuations and measurement errors are aleatoric uncertainties. They are usually treated with statistical means.

In contrast, epistemic uncertainties account for systematic deviations that could in principal be identified, but which are not considered for practical reasons. For instance whenever simplified models are used, neglecting certain aspects of a more complex model, the loss of information due to this inferior model allocates to epistemic uncertainty.

There are various works related to including uncertainty into machine learning algorithms, where Shaker and Huellermeier [1], considering a random forest approach or Depeweg [2], focusing on Bayesian neural networks, are only two examples.

Past works also showed the advantages of introducing additional knowledge into the machine learning training, forcing for instance the latent layers of a neural network into a fixed shape. Bishop showed in [3] that networks could be reconfigured to forecast also probability distribution moments. Closely related to this approach is the concept of the variational autoencoder (VAE) [4], which constructs the latent layer from the moments of a probability distribution. Autoencoders are known for the application in fault detection. Gao et al. [5] and Wang et al. [6] show examples that apply a VAE to complex problems. In Srikotr and Mano [7], the authors demonstrate the quantisation of spectral envelopes with VAE. For realizing effective intrusion detection, Xu et al. [8] deploy specifically conditioned VAE for the pattern surveillance.

From these "design" ideas for neural networks, the clear next step is to introduce physical relations into the network topology, a concept called physics-informed (sometimes also physics-infused or physics-qualified) machine learning. Here, the advantages of traditional first-principle modelling are combined with the powerful possibilities of the learning algorithms, either to solve analytical problems like differential equations as presented by Raissi [9] or to apply physical knowledge within the network, which was depicted by Shukla et al. in [10].

Defining enriched input data for training a neural network is a specific subtype of physics-informed training. Here, prior knowledge (from domain expertise) about the best preparatorial transformation is used for feeding the network with right information. One might argue, that this is contradictory to the actual idea of machine learning, because it suppresses or shortcuts a mayor advantage of learning: the automation of information retrieval. Yet, for practical reasons such prior knowledge is helpful. It reduces the training times and renders complex machine learning results interpretable in a technical sense.

In the current work, we will consider the application and integration of uncertainties into neural networks for forecasting anomalies and process outcomes of production processes. We will investigate how variational autoencoders [4] can be utilized on process data with challenging anomalies. In this context, we will also analyze the distribution functions in the latent encoded space, which are highly effective for determining process perturbations.

The intention of this work is not to introduce a new way of anomaly detection, but rather to discuss, how developers of anomaly detection systems can utilize and adapt existing techniques like autoencoding for their needs. The chosen example is simple enough to illustrate the methods and the impact of autoencoding on the latent space. There are many known solutions, like deep neural network classifiers or straightforward Fourier transformation to actually detect the anomaly equally good in the given case, but most of them require a prior knowledge about the shape of the perturbation.

2 Stochasticity in Production Processes

2.1 Analytical Treatment of Fluctuations

The concepts discussed above are also relevant for a more theoretical reason: beneath machine learning approaches, the correct identification of any involved stochastic process (variable) is of great importance for analytical models, too.

Including stochasticity for describing multi-particle motion and transport phenomena like diffusion has a longstanding history, reflected clearly in the widely used Langevin and Fokker-Planck formalisms [11,12].

Assume the analytical evolution of the process is known and described by the function $\tilde{x}(t)$, then the superposition of this function with a stochastic process ν that models the fluctuations leads to

$$x(t) = \tilde{x}(t) + \nu, \tag{1}$$

where the stochastic properties of ν are determined by the moments of the associated probability distribution

$$\mu = \langle \nu \rangle = \int_{-\infty}^{\infty} \nu P(\nu) d\nu \tag{2}$$

and

$$\sigma^2 = \langle \nu^2 \rangle - \mu^2. \tag{3}$$

Thus, the problem transforms into extracting the probability function $P(\nu)$.

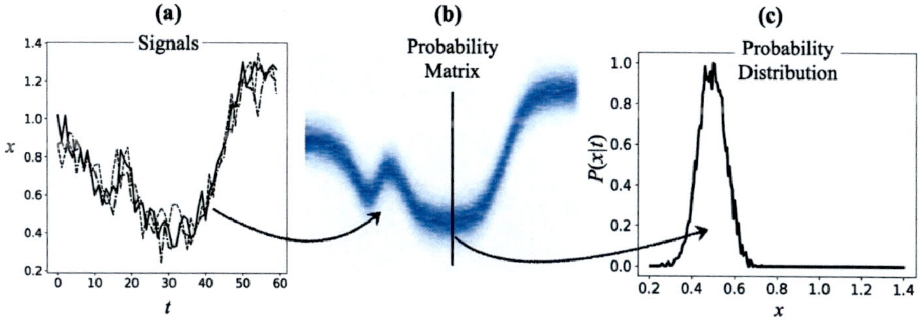

Fig. 1. Illustration of the probability matrix (corridor) build from a group of signals. Assorted example signals are shown in (a), the matrix is visualized in (b). A slice of this matrix is plotted in (c), yielding a near-gaussian distribution.

2.2 Example and Its Perturbation

Figure 1(a) shows a series of normal process curves, where our example is chosen from a set of industrial engine voltages that have been properly rescaled. The production is highly repetitive and we may safely assume that the signal should behave the same. There are several known anomalous events in this signal, but the effect that is most difficult to detect is oscillatory perturbation in the voltage.

In Fig. 2 this perturbation is shown in more detail, yielding also the difference to the normal case. Note that this perturbation has several features of interest: i) it is bounded well within an envelope around the signal, ii) it is difficult to detect it in a series of processes, even by the expert eye and iii) although there is only a single frequency displayed in Fig. 2, it may feature other frequencies as well.

Formally, we define $\mathbf{G}$ as the set of all normal (good) situations and $\mathbf{B}$ as the set of all bad scenarios. Of course for our example we can also further divide these sets into $\mathbf{G}_{\mathbf{Train}}$, $\mathbf{G}_{\mathbf{Test}}$, $\mathbf{B}_{\mathbf{Train}}$ and $\mathbf{B}_{\mathbf{Test}}$.

2.3 Classical Approach for Extracting Uncertainties

We continue our considerations by adopting a very simplified perspective. First we would like to properly describe our uncertainty corridor along the signal. To do this, we let $x \in X$ be a stochastic process, namely some time series with discrete time steps describing a process variable of a production process. Then $P(x|t_j)$ with $j \in \mathbb{N}$ describes the probability of finding the value x at a given time t_j. The set X can be further discretized by a limited number of intervals $\xi_i = [\xi_{i_<}, \xi_{i_>}]$ such that with

$$\xi_{i_<} < x < \xi_{i_>} \Rightarrow x \in \xi_i \tag{4}$$

a probability matrix $\mathbf{P} = (P_{i,j})$ can be defined by

$$P_{i,j} = P(x \in \xi_i | t_j), \tag{5}$$

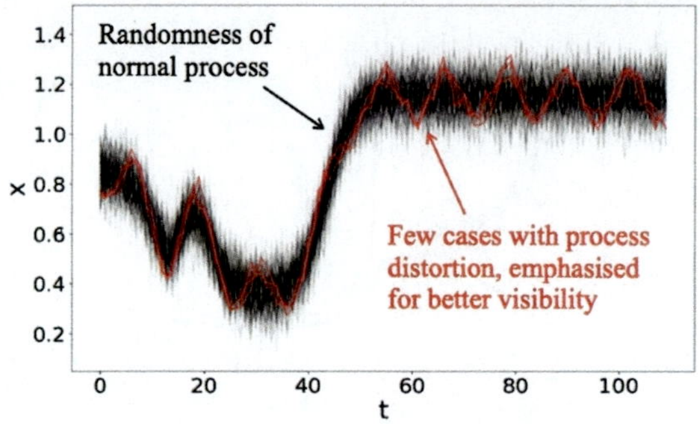

Fig. 2. The anomaly visualized: black is the normal process situation, in red, distortions of the process are emphasized. The distortion is well bounded by the process noise and difficult to find by statistical approaches.

where the size of **P** depends on the number of intervals ξ_i and the number of time steps t_j. Please note the important difference to $P(\nu)$ in (2) that now **P** includes the overall analytical motion $\tilde{x}(t)$, more explicitly the path described by $\tilde{x}(t)$ is the path of the distribution average of **P**.

Once (5) is found, it is already a convenient tool to detect anomalies in the process, as any encountered x in an interval ξ_i with a low probability, would clearly point towards an anomalous behaviour of the process. Figure 1 illustrates this by assuming a process time series in Fig. 1(a). Using a sufficiently large data set, the matrix **P** can be trained to reflect the statistical behaviour of the time series as presented in Fig. 1(b), where the intensities scale with the numbers of the matrix. Finally, the column vectors of the matrix **P** at t_j represent, of course, the individual conditional probability distributions.

2.4 Limitations

The shown method or very similar variants of it, are often found in industry for heuristic anomaly detection. It only requires the knowledge of some statistical key figures and an estimate of the shape of the normal process curve. Although this probability corridor is a good approach for observing deviations from the path, it has several shortcomings:

1. It does not detect deviations that stay within the statistically defined normal path, even if the principal form is significantly distorted. This is for instance the case, when a sinusoidal noise perturbs the original signal, with an amplitude small enough to be covered in the noise fluctuation.
2. The assumption that the distribution $P(x|t_i)$ is independent of the prior position $x(t_{i-1})$ is a mayor simplification. It implies that for any position $x(t_{i-1})$ the subsequent probability distribution for the next time step is equal.
3. Any continuous deterioration over time that is used within this approach will just be reflected in a broadening of the probability corridor and thus not properly account for the presence of an anomaly.

3 Autoencoders and Uncertainty

3.1 Standard Autoencoder Concept

An autoencoder is a neural network, which feeds the same set of training data into both, the input layer and the output layer. Consequently, it forces a transformation of the input data onto itself, namely a unity mapping. As there are no labels principally required, autoencoders are unsupervised learning algorithms. The hidden layers of the topology are symmetric and subsequently decreasing in the number of neurons. When mapping the information from input to output, it has to pass a bottleneck with only few numbers describing the actual signal. Several activation functions can be used for building the network and common back-propagation with an adequate cost function will provide good autoencoding results.

Figure 3 shows an illustration of an autoencoder network. The latent space neurons are marked in red. The yellow neurons represent the encoder part and the blue neurons are the decoder part.

Autoencoders are therefore dimensionality reducing network paradigms - where the reduction is generated by the constriction topology in the hidden layers. In contrast to other dimensionality reduction approaches like the (linear) principal component analysis, autoencoders feature a highly nonlinear transformation and can thus be applied to more complex problems.

A clear difficulty of common autoencoders is the interpretation between input and encoded input, as the latent space representation neither must be continuous nor does it support simple interpolation. Even small changes of the latent neurons might produce strong differences in the output.

3.2 Variational Autoencoders for Dimensionality Reduction

These problems can be overcome, when we introduce constraints for the latent space. Variational autoencoders [13,14], enforce a mapping of the input data directly into a probability distribution, more explicitly into its moments. Assume we have k latent neurons representing the k moments m_k of a probability distribution $P(x, m_1, m_2, \dots)$,

$$m_k = \int_{-\infty}^{\infty} x^k P(x, m_1, m_2, \dots) dx. \tag{6}$$

In practise, one restricts to the lower moments like mean $\mu = m_1$ and variance $\sigma^2 = m_2 - m_1^2$ shown in (2) and (3). For enforcing now a given distribution Q, the cost (or loss) function is extended to allow for minimizing the Kullback-Leibler divergence,

$$\Delta_{\mathrm{KL}}(P||Q) = \sum_x P(x) \log \frac{P(x)}{Q(x)} \tag{7}$$

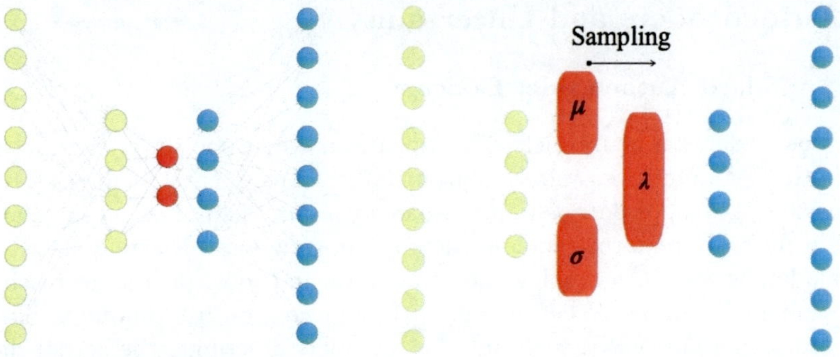

(a) Standard Autoencoder (b) Variational Autoencoder

Fig. 3. Illustration of two types of autoencoders.

between this reference distribution $Q(x)$ and the achieved training iteration of the probability distribution $P(x)$. Gaussian normal distributions $\mathcal{N}(\mu, \sigma^2)$ or Bernoulli distributions are frequently used for this type of encoding, but also hyperspherical [15] and Gaussian mixtures [16] have been applied successfully.

Construction of a variational autoencoder requires also the so-called reparametrization "trick" introduced by [17], that ensures we can train the network with the backpropagation algorithm. To understand this reparametrization, we start with the forward evaluation, where it is necessary to perform the step from the latent space into the first decoder layer. This is done by sampling from the distribution defined by the latent space neurons, thus creating a latent space sample vector λ that has now to be propagated through the decoding part of the network,

$$\lambda = \mu + \sigma\nu \tag{8}$$

where μ and σ are deterministic quantities and ν is a stochastic process, resembling the noise perturbation shown in (1). ν is here the only random quantity.

3.3 Variational Autoencoder with Physics-Informed Input Modification

We are now interested in investigating the results of the autoencoder to different transformations of input data. Therefore, we introduce a transformation for $x(t)$ in terms of

$$L[x(t), \mathbf{w}] = \sum_{i=0}^{N} w_i \frac{d^i}{dt^i} x(t) \tag{9}$$

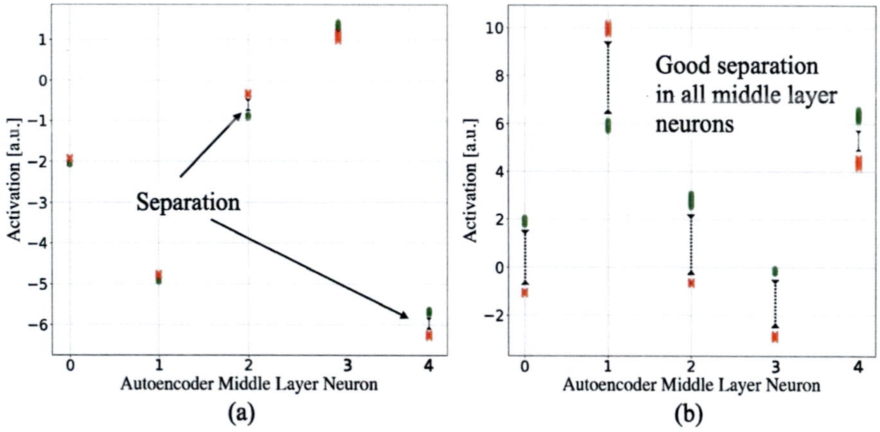

Fig. 4. Results of shallow autoencoder transformations on challenging anomaly scenario: (a) a 500/50/5/50/500 topology transforming the input signal onto 5 bins and (b) the result of a physics-informed 500/50/5/50/500 autoencoder network. (Color figure online)

which can be seen as a weighted sum of derivatives of $x(t)$. With the training data in $\mathbf{G}$ and $\mathbf{B}$ at hand, we can find the weights by solving following optimization problem,

$$\begin{aligned} \underset{w_i}{\text{maximize}} \quad & f_{\text{phys}} = \sum_k |\xi_k - \zeta_k| \\ \text{subject to} \quad & \text{(C1)}\ \xi_k = L[x_k(t), \mathbf{w}] \ \text{ with } \ x_k \in \mathbf{G} \\ & \text{(C2)}\ \zeta_k = L[z_k(t), \mathbf{w}] \ \text{ with } \ z_k \in \mathbf{B} \\ & \text{(C3)}\ w_i \geq 0 \end{aligned} \tag{10}$$

So far, we have seen that we can enhance the evaluation of the autoencoder by a simple transformation, but this comes at a cost: we require prior knowledge of the anomaly case situation, which is usually available when (like in the above training) the training set included explicitly a contribution of the anomalies. It renders the overall approach to be a supervised learning technique, as again labels for the data are necessary.

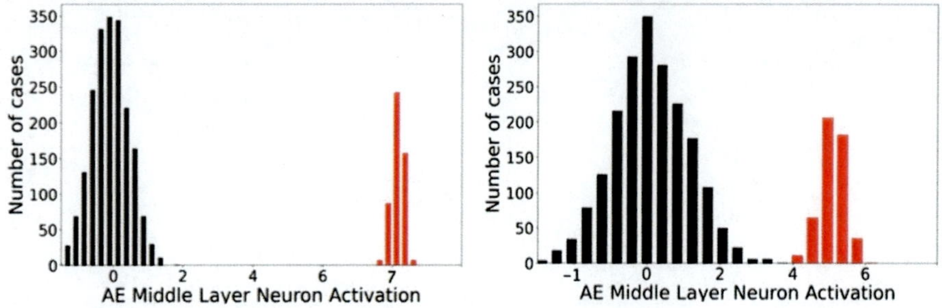

Fig. 5. Distribution of the latent space variables, normalized in x direction and calibrated, so that the good cases are centered at $\hat{\mu} = 0$. Left: Training with 25% of bad cases included in the training data set. Right: Training without any bad cases represented in the training set.

4 Interpretation of Resulting Distributions

4.1 Clustering in the Autoencoder Latent Space

Figure 4(a) shows the results of an autoencoder with a topology of 500, 50, 5, 50, and again 500 neurons. The figure can be understood as follows: we plot the activation value of each neuron in the latent layer of the autoencoder against its position. The middle layer contains the transformed and most compressed state of the data. Then, we evaluate dedicated test sets with good cases (green dots) and bad cases (red crosses). If the clusters which represent the bad cases do not mix with the good cases, the autoencoder was able to distinguish these scenarios successfully. At least two latent layer neurons react to the distortions in our situations, neuron 2 and 4. With their values, a separation between the good and bad cases can be established. It is indeed sufficient for an anomaly announcement.

In comparison with Fig. 4(a), the results of the physics-informed variational autoencoder being trained with the optimized set of data is shown in Fig. 4(b). The separability is significantly improved by the added prior knowledge. The separation is now visible in all neurons of the latent layer. The distances between the case clusters have also been increased by the transformation.

4.2 Probability Distribution in the Latent Space

The clusters we found in 4.1 correspond to distributions - now in latent space - which relate a test case with the probability of belonging into the categories good or bad. We can consequently evaluate the clusters in form of histograms.

A variational, physics-informed autoencoder with a network topology of 500, 100, 100, 10, 5, 10, 100, 500 was now trained (without cross-validation) in two distinct situations: a) with the training data set contained approximately 25% of the anomalous situations and b) with a training set that has no single anomalous case. Note, for the physics-informed transformation, an equally distributed set of w_i was assumed and not optimised, as we fully assume the anomaly is unknown before training. The test set was composed of 2500 situations with 20% of bad cases. Following conclusions can be drawn:

- Both trained algorithms found a good separation between the good and the bad cases. As expected the quality of the differentiation in terms of probability distributions in the latent variables is better if the anomaly is known in advance. Figure 5 (Left) shows the result of the algorithm that was trained with 25% of anomalous situations. Its distribution of good cases was rescaled and centered around 0. The smaller, red distribution represents the bad cases, which are well separated.
- In Fig. 5 (Right) we see the algorithm results when the training did not contain any anomalies at all. The same scaling was applied than for the algorithm results on the left. As expected, the distributions are much closer to each other. The differentiation between good and bad is still good, but apparently more difficult than in the other case. Also the variance of both the distribution of good cases and the distribution of bad cases increased.

5 Conclusion and Outlook

The paper demonstrated the application of autoencoders to the problem of anomaly detection. It investigated how uncertainty can be included into the treatment and how it affects the output results. For the autoencoder, beneath an adequate constriction ratio, it is important to constraint the latent space itself. Variational autoencoders are a good example for this, as their latent space is reproducible. The additional use of physics-informed transformations and additional data makes the procedure more sensitive, but leads to the drawback, that the anomaly situation must be known at the training time.

Tests with training on data that even excludes explicit knowledge about the anomalies, as they are shown in Sect. 4, indicate that the autoencoder learned what a normal process situation is and was capable to distinguish the deviation from the normal situation very well. The anomaly in this case was oscillatory. It was well enclosed in the process stochastics and was detectable by other minimum or maximum conditioning techniques. A Fourier transformation would have detected the anomaly, but this would also need prior information on the anomaly, before it occurs.

The advantage of the autoencoder is therefore its capability to find principally unknown anomalies with a satisfying quality. What is the advantage to incorporate the uncertainties? It straightforwardly helps to assess limiting corridors of industrial production processes. Knowing the latent distribution also helps to determine optimal thresholds for the anomaly detection.

References

1. Shaker, M., Huellermeier, E.: Advances in Intelligent Data Analysis XVIII. ch. Springer, Aleatoric and Epistemic Uncertainty with Random Forests (2020)
2. Depeweg, S.: Modeling epistemic and aleatoric uncertainty with bayesian neural networks and latent variables. PhD thesis, TU Muenchen (2019)
3. Bishop, C.M.: Mixture density networks. tech. rep. (1994)
4. Kingma, D.P., Welling, M.: An introduction to variational autoencoders. Found. Trends Mach. Learn. **12**(4), 307–392 (2019)
5. Gao, Y., Shi, B., Dong, B., Chen, Y., Mi, L., Huang, Z., Shi, Y.: Rvae-abfa: robust anomaly detection for highdimensional data using variational autoencoder. In: IEEE 44th Annual Computers, Software, and Applications Conference (COMPSAC), pp. 334–339 (2020)
6. Wang, K., Chen, J., Song, Z.: Fault detection based on variational autoencoders for complex nonlinear processes. In: 12th Asian Control Conference (ASCC), pp. 1352–1357 (2019)
7. Srikotr, T., Mano, K.: Predictive vector quantized variational autoencoder for spectral envelope quantization. In: International Conference on Electronics, Information, and Communication (ICEIC), pp. 1–4 (2020)
8. Xu, X., Li, J., Yang, Y., Shen, F.: Towards effective intrusion detection using log-cosh conditional variational autoencoder. IEEE Internet Things J. p. 1 (2020)
9. Raissi, M., Perdikaris, P., Karniadakis, G.E.: Physics-informed neural networks: a deep learning framework for solving forward and inverse problems involving nonlinear partial differential equations. J. Comput. Phys. **378**(1), 686–707 (2019)
10. Shukla, K., Di Leoni, P.C., Blackshire, J., Sparkman, D., Karniadakis, G.E.: Physics-informed neural network for ultrasound nondestructive quantification of surface breaking cracks. Journal of Nondestructive Evaluation 39, Art. No. 61 (2020)
11. Kampen, N.G.V.: Stochastic Processes in Physics and Chemistry. Elsevier (2011)
12. Gardiner, C.: Stochastic Methods. Springer, A handbook for the natural and social sciences (2009)
13. Bishop, C.: Pattern recognition and machine learning. Springer (2006)
14. Shafkat, I.: Intuitively understanding variational autoencoders. http://towardsdatascience.com last checked: 2020-10-30 (2018)
15. Davidson, T.R., Falorsi, L., De Cao, N., Kipf, T., Tomczak, J.M.: Hyperspherical variational autoencoders. In: Conference on Uncertainty in Aritificial Intelligence (UAI). Monterey, CA, USA (2018)
16. Dilokthanakul, N., Mediano, P.A., Garnelo, M., Lee, M.C., Salimbeni, H., Arulkumaran, K., Shanahan, M.: Deep unsupervised clustering with gaussian mixture variational autoencoders. arXiv preprint:1611.02648 (2016)
17. Kingma, D.P., Welling, M.: Auto-encoding variational bayes. In: Conference Proceedings of the 2nd International Conference on Learning Representations (ICLR) (2014)

Mapping of Standardized State Machines to Utilize Machine Learning Models in Process Control Environments

Julian Grothoff(✉) and Tobias Kleinert

RWTH Aachen University, Chair of Process Control Engineering, Turmstr. 46, 52064 Aachen, Germany
{j.grothoff, kleinert}@plt.rwth-aachen.de

Abstract. Digitalization is a driving force for new applications of artificial intelligence in the industrial context. In the last decade new and optimized machine learning methods, tools and corresponding Hardware became available. But their utilization in industrial applications is very challenging, due to different requirements and conditions. In order to bridge that gap, trained machine learning algorithms need to be integrated in current control system environments, which implies a setup of the application program based on components of similar interfaces and a standardized behavior. Hence, a mapping of well-known and broadly applied state machines, such as the ANSI/ISA-88 (also known as PackML) states and modes for batch control, to generic interactions of reinforcement learning algorithms is presented in this contribution. Additionally, a simplistic proof of concept using reinforcement learning in a simulation of a coil transport system for cold rolling steel mills is shown. Moreover, the interaction of a reinforcement learning agent with its standardized state machines is described for learning and interference phases.

Keywords: Reinforcement learning · Process control components · PackML state machine

1 Introduction

Artificial Intelligence (AI) and Machine Learning (ML) are growing research fields. Hence, new methods and application possibilities open up. For several years, "Industrie 4.0" and other comparable initiatives have been reaching out to digitalize industrial production. Central targets are to develop data handling systems and process control concepts which allow for the industrial utilization of AI and ML.

Nowadays, standardized and reliable control systems ranging from the manufacturing execution system (MES) to programmable logic controllers (PLCs) are used that are, however, to a certain extend inflexible with respect to the implementation of more advanced data processing and control concepts. These control systems are very different compared to AI training environments: Coding relies on e.g., IEC-61131 programming languages or C/C++ whereas in the area of AI and ML, Python, "R", AI frameworks (e.g. Tensorflow, PyTorch, …) and others are widely used. Moreover,

V. Colla and C. Pietrosanti (Eds.): ESTEP 2020, AISC 1338, pp. 39–53, 2021.
https://doi.org/10.1007/978-3-030-69367-1_4

industrial applications often comprise different requirements in terms of real-time capabilities, availability, long term support, etc., that need to be considered.

In order to guarantee reliable process operation, control systems which incorporate AI-applications should be designed and operated in a unified manner, while taking existing standards into account. In consequence, an operator should see no difference in how to handle and execute an AI-powered or a "regular" control unit.

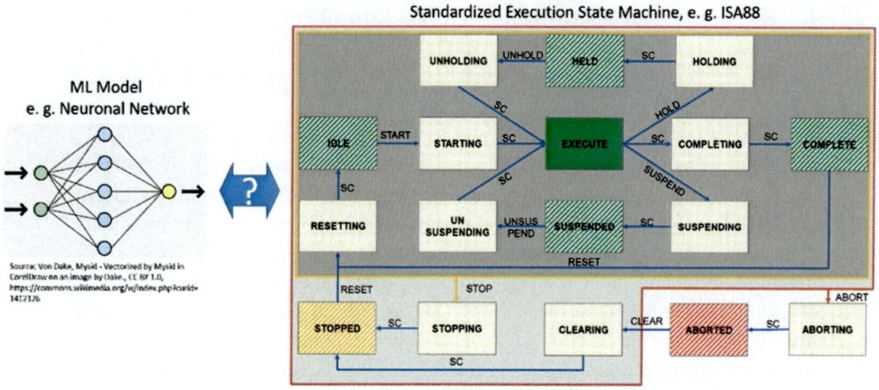

Fig. 1. Exemplary visualization of the mapping of ML control states and the states of an automation system control component

We propose a PLC-compliant, component-based integration of ML algorithms into process control systems. The key aspect is to map broadly established, standardized execution states and modes of process control components, e.g. defined by ISA 88 or PackML standard (c.f. Fig. 1 right), to generalized states of ML algorithms in order to create subsystems of process control systems that are capable to handle AI and ML reliably in an industrial control context. Our goal is to address unified integration and operation of ML algorithms in Operational Technology (OT).

2 Utilization of Machine Learning in Control Applications

Machine learning is increasingly used to find errors in production processes or to determine product quality in industrial applications. Hence, it only accesses data or sensors and has an indirect impact on the process control system as well as its actuators. In contrast, in this contribution we focus on the replacement of control functionalities, like feedback or sequence control, inside the process control hierarchy. The approach is motivated by the need to adapt fast to unplanned product type and production condition changes due to increased individualization and dynamic global markets. Furthermore, the (online) optimization of multi parameter or complicated objective functions may be solved with ML methods in direct control in the future.

2.1 Control Applications in the Context of Industry 4.0

At first, we discuss related work utilizing machine learning for control applications. ML and reinforcement learning (RL) in particular is often used in robotics for dynamic handling operations, as these are hard to implement programmatically [1, 2]. Other recent use cases consider the load distribution in power systems [3, 4] or (nonlinear) model predictive control (MPC) in process automation [5, 6] for example. Even some benchmarks in the field of technical process control for feedback controller can be found [7]. Other use cases of RL involve scheduling [8] or decision-making processes [9] above the process control system (PCS) layer or on existing control functions [10]. These implementations are often coupled with agent-based concepts [11]. Contribution [12] presents a deployment approach for RL agents using a PLC as a training environment. An overview of different ML and optimization approaches is covered in [13]. In conclusion, the utilization of ML and RL in specific use cases and with specific methods is a popular ongoing research topic. In contrast this contribution is focussed on the generalized integration of ML methods into (brownfield) production environments, combining current modularization efforts for service-based production with RL.

The current research and standardization initiatives in the field of digitalization, e.g. "Industry 4.0", "Digital Factory", "Web of things", will lead to new software architectures in production. Compare IEC TR 63319 ED1 (currently in APUB status) for an international overview of meta models in development. These architectures and corresponding technologies (e.g. Asset Administration Shell, OPC Unified Architecture) will support the concepts of this contribution by delivering a higher connectivity, comprising more data and simulation models available. Moreover, a higher adaptability of production environments will be achieved and requested. The german project BaSys 4.0 and its follow up BaSys 4.2 (https://basys40.de) are working on a realization of Industry 4.0 architectures. The control component concepts that are utilized for this contribution, were developed in their context. The project ensures compatibility to other similar efforts, like NAMUR Modular Type Package, Open Process Automation Standard, OMAC Packaging Machine Language (PackML) or different OPC UA Companion Specifications.

2.2 Reinforcement Learning for Plant Control

The traditional approach to design a control function is to write a program or algorithm that calculates output values for given input values. The algorithms and their implementations are often derived from analytical or empirical models of a process. In the automation domain the logic is implemented in specialized programming languages such as structured text, ladder diagrams, function blocks and continuous or sequential function charts according to IEC-61131-3 or IEC-61499.

Compared to the traditional approach, the ML approach is divided into two phases: (1) the training or offline phase, which takes input and output data and adapts (learns) the structure and parameters of a model. This model could be e.g., an artificial neuronal network or a state vector machine. (2) In the second phase, the model is used as a program. This "inference" phase is analogue to the traditional approach, but usually other input data is used – e.g. real-time process data instead of recorded training or test data sets. Under certain circumstances it is even possible to mix both phases and data sets, called online learning. The important difference compared to the classical control design is that relevant input and corresponding output data needs to be provided instead of the algorithms or the control logic itself.

In practice it may be easier to simulate or use an environment that realizes the controlled system instead of generating or gathering input and output data for ML. This is the basis for reinforcement learning (RL). Usually RL is carried out as supervised learning, where the supervisory part specifies the rewards – the reinforcement, also called objection or cost function – based on observations of the environment. The observations of the supervisor (trainer) may be the same or different than the environment. In the context of RL, the trained relation between input and output data is named policy, as the input data corresponds to observations and the output data to actions used to affect the environment. Moreover, the training (also called learning algorithm) and the policy are combined as an agent. In some realizations, the supervisor is depicted as part of the environment which is implicitly true for environments that expose the value of a cost function, like a key performance index (e.g. product count or quality measures) in manufacturing execution systems.

In conclusion, a proper simulated or protected environment, as well as a set of reward definitions, need to be provided in order to automatically learn a control function or strategy with RL instead of engineering it. Obviously, it is not easy to implement, optimize or generalize the learning process. It is even possible that some specific control behavior cannot be learned at all. This may lead to an RL agent that performs wrongly or worse than a properly engineered solution. In this paper we focus on the integration of RL models in the industrial (OT) context and consider performance design as part of a data-based process optimization activity instead. Hence, the performance optimization is assumed to be solvable by available or future AI methods. However, we will apply RL in the considered example of a palette transport system as a proof of concept.

2.3 Component Based Process Control

In order to achieve a high adaptability of the control system to unplanned changes of the controlled system (the plant), it is reasonable to use a component-based approach. The modularization of control logic in combination with a service-based control architecture makes it possible to easily change, replace and add control functionality, as well as scale and maintain it [14].

In industrial IEC-61131-based environments the control logic is often realized in PLCs, which are aggregated and monitored by a centralized process control system (PCS). In practice the technical functionality is typically implemented in a complex mesh of function blocks or as monolithic application in contrast. A migration path based on these preconditions is shown in Fig. 2.

At first the elementary, standardized and industry-proven function blocks from various libraries can be reused by separation of general and specific parts of function block networks. The more general and application agnostic parts can be mapped to standardized state machines. These state machines can be encapsulated in a component, called "Control Component" (CC) [15]. By doing so, the encapsulation has to

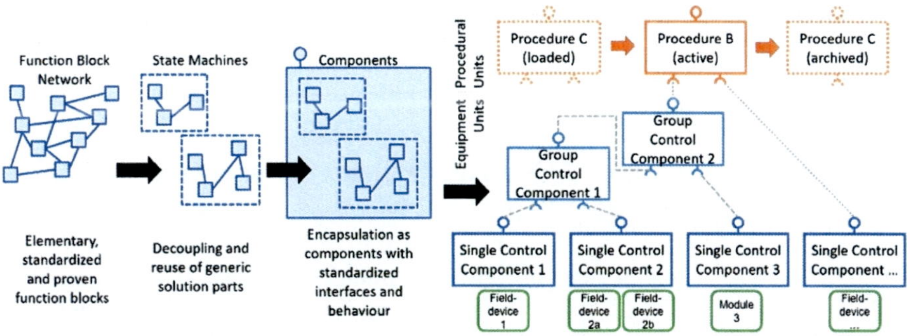

Fig. 2. Migration from function block networks to a component-based control application

guarantee the handling of the CC as a single, compact unit. The technological and application specific parts of the control logic are called operation modes, which can be considered as selectable services of a CC. To sum up, control logic can be transformed to unified CCs comprising the following most relevant state machines:

- Occupation State Machine: the occupation determines **who** controls the component, i.e. the operator, another CC, a procedure or no one at the moment.
- Operation Mode Selection: the current operation mode specifies **what** the component does, as it encapsulates the application specific control logic.
- Execution Mode Selection: the execution mode influences the execution of an operation mode. It determines **how** the execution (control flow) is carried out. Typical modes are automatic, semi-automatic, manual. These modes may also be called production mode, maintenance mode etc. [15] Often the execution mode affects the consideration of speed limits, acknowledge and safety signals.
- Execution State Machine: the execution state determines the **activity** of the whole component and the active operation mode. Typical execution commands are start, stop, reset, etc. (C.f. Fig. 1 right)

These state machines are designed to be nearly orthogonal to each other and to describe a variety of the CC's current conditions in an abstract way. They can be found in different standards in specialized, collapsed or combined forms [16]. Examples which are used later, are the "states" and "modes" defined in IEC-61512, also known as ANSI/ISA-(S)88 or as PackML(ISA-TR88.00.02). Moreover, components of different complexity levels can be realized, so that not all state machines and states are needed. To overcome the different implementations, their variation and mapping to the abstract state machines can be described via so called profiles [17].

Based on the unified behavior of CCs a simple service interface can be defined [15]. The interface may be implemented utilizing different communication technologies, e.g. OPC-UA, MQTT, HTTP/REST. The interface may even be realized with a human-readable string-based command interface in PLC environments.

Hence, it is advantageous to realize the whole control system from field devices through PCS up to the MES as a hierarchy of CCs, as depicted in Fig. 2 on the right. In this paper we focus on group CCs, which represent higher-level equipment-related functionality. The concept also applies to single CCs to execute the control of field devices or smaller modules. Furthermore, CCs can be used in orchestration scenarios, e.g. in product or process related procedures.

3 Scenario - Learning to Control a Palette Transport System of a Cold Rolling Mill Plant

To show the interaction of a reinforcement learning agent with standardized state machines and the states of subsystems which it controls, a simplistic proof of concept is shown. The palette transport system (PTS) for coils of an industrial real-time simulation forming an aluminum cold rolling mill (CRM) plant was rebuilt in a simplified manner using the Unity 3D engine. The behavior of CCs was implemented for individual conveyors and for the prototype of an AI-powered group control component that represents a palette for coils. The Unity ML-Agents Toolkit [18] was used with a proximal policy optimization from the Tensorflow framework to perform the coils logistics.

3.1 Migration from the Real Plant to the Gym

In order to get from the real plant to the training environment (also called gym or academy) the following environments were used, as shown in Fig. 3 from right to left.

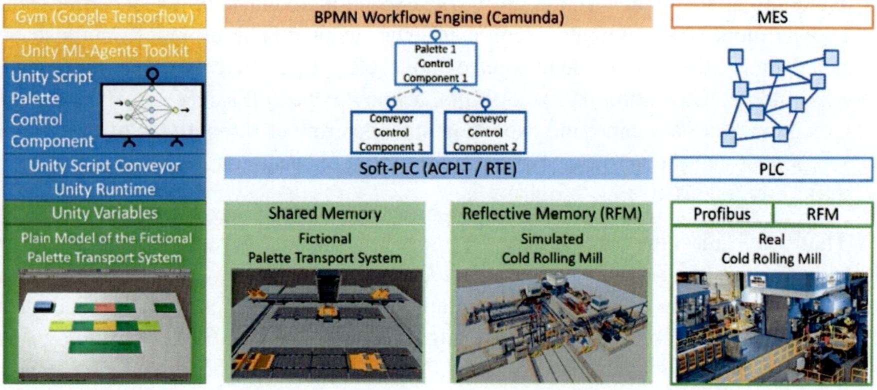

Fig. 3. Environments and software stacks for the coil transport system scenario.

1. Real CRM: the field devices of the real plant are interconnected with classical field busses like Profibus and with a reflective memory (RFM). Via this (remote) I/O layer traditional PLC programs execute orders from the MES system. As the component-based control approach is state of the research but not yet state of the art, testing and even training on the real plant could not be carried out so far.
2. Simulated CRM: SMS Group GmbH simulates their plant automation system in complex integration tests before commissioning with a high level of detail in an

industrial real-time simulation. In the context of the BaSys 4.0 and BaSys 4.2 projects (http://basys40.de) such a simulation was provided to the authors and 14 other project partners. Parts of the simulated CRM were automated to evaluate and show the concept of CCs via an open-source runtime environment (ACPLT/RTE) as Soft-PLC in an IEC-61131 manner.

3. Fictional PTS: in order to demonstrate interoperability and adaptability of the CC concept, the PTS was separated as a subsystem of the simulated CRM with a reduced shared memory (SHM) access. The conveyors were automated by different project partners with different technologies independently, e.g. IEC-61131 FBs, IEC-61499 FBs, C/C++ code, Java code, OPC UA and HTTP/REST. Afterwards a group CC was able to navigate a palette through the PTS and CRM on shiftable and turnable conveyors.
4. ML environment for the PTS: the simulated PTS or CRM could be used for training purposes directly, but in order to use the integration of the Unity ML-Agents toolkit and for parallel, time-lapsed training a plain model of the PTS was created. This allows to focus on necessary parts for training and for the validation in the PTS or CRM simulation, treating them as the real plant.

3.2 The Prototypical Gym

The technological goal for the prototype (Fig. 3. left) was to create a ML operation mode in a CC representing a palette in the PTS utilizing RL. The first simple task for the RL agent was to learn, how to get to the end position in the logistic system by shifting two conveyors from random positions to the right track and controlling the current and next conveyor. This service (operation mode) of the palette CC is already higher-level compared to normal operation modes, e.g. left, up, turn and heat.

The prototype [19] was realized completely in the Unity software. For this reason, the training environment was realized with simple cube objects, representing conveyors and palettes, which were extended with C# scripts for the dynamic behavior. The conveyor objects comprise a motor object with a script attached that utilizes Unity's physics engine to move palettes on top of them. Additionally, four light barrier objects with simple Raycasts were added detecting the palette positions like in the PTS simulation and in the real plant. The conveyor objects were equipped with a script inheriting from "IControlComponent" interface, which was derived from the CC concept. Up to now, only execution states, the mode selection and occupation were realized. The conveyors mimic the CCs behavior, implemented on top of the PTS and CRM simulation. The execution of operation modes was implemented using Unity's concept of "Co-Routines", mapping the control flow to a sub state machine of the unified execution state machine. The control of the palette object was realized with a C# script which inherits the functionality of the "agent" class of the ML-Agents library [18] and the IControlComponent interface, combining agent and operation mode.

3.3 ML Operation Mode Enhancement

The task of reaching the end position can be enhanced to deliver a service of even higher level that could be trained via curriculum learning. Such enhancements include the dynamic specification of the end position, whether the palette should be turned on its way, how to deal with multiple palettes, blockades and structural changes to the conveyor topology. This demonstrates the promising benefits of an RL-CC-based application program, namely to automatically cope with complex tasks and dynamic changes.

The PTS also includes an oven to heat the coil. Here we can emphasize the difference of traditional engineering, where we would control a target temperature of the oven, whereas for the ML operation mode, we could ask for the coil temperature at the end position, taking transportation times and properties of the coil into account, like its dimension, by rewarding the agent for a lower temperature difference. This behavior would be hard to engineer in a classical approach, taking all possible events into account. Another advantage is that the palette can nearly start in any position and state (even if it wasn't trained).

In conclusion, the plain model can be used to experiment with and optimize the training process, e.g. for hyper parameter tuning and reward shaping. The PTS model can be used for evaluation, in analogy to a validation data set in classical machine learning, however, in a much broader sense. Furthermore, it can be used for software in the loop tests (inference), embedding the ML model in PLC code. The CRM model can be used for tests of generalization, adaptability and scalability of the learning procedure and the model's policy, as well as for the design of an automatable ML integration workflow.

4 Interaction Scheme of the Machine Learning Model and Process Control Components

An ML model or rather a RL agent could be implemented inside CCs in different ways. The deployment could be internal (c.f. [12]) or external e.g., via network communication. In the prototype it is realized internal, but in a PLC runtime it could be necessary to place the ML policy external e.g., into an edge cloud. Both solutions are possible with different impacts, like communication or computation overhead. Moreover, dedicated hardware is available for running such models in inference.

The following alternatives may be considered, regarding the available model or policy actions. First of all, we assume an integration inside an operation mode. But the operation mode could be used to select from other existing operation modes inside the CC. For example, in our scenario the agent could select from the control commands "left", "right", "up", "down", "turn" and "heat". In general, it may even select these operation modes in parallel. The advantage is that the model uses tested and comprehensible (engineered) logics (c.f. explainable AI). The opposite approach is to let the model choose all available commands for each command output (connection to underlying CCs or field devices), which creates a huge solution space with more degrees of freedom. In the scenario this would let the agent pick from commands for occupation (occupy, free), execution (start, stop, reset, …, automatic, manual, …), as well as operation modes (forward, backward, shift up, …) for each of its four command outputs corresponding to the current, next and the two shift conveyors.

More solutions are possible in between both extremes. In general, we want the agent to have as many degrees of freedom, as it needs to fulfill its task in an optimal way. But we only want it to have technologically reasonable choices. A generic solution is to let the agent choose an operation mode for each command output, which can be achieved by creating macros for them. These macros can be generic, as they follow the same sequence of occupying, selecting the desired operation mode, sending start, waiting for the execution to complete and freeing the underlying CC. This sequence is based on the unified state machines and interfaces of CCs. Moreover, the agent doesn't need to learn the sequence and prevents it from making mistakes, like trying to start without occupation or selected operation mode. In consequence, the command output macro solution was implemented in the prototype as "ExecuteOpMode" function.

4.1 ML Model Interactions

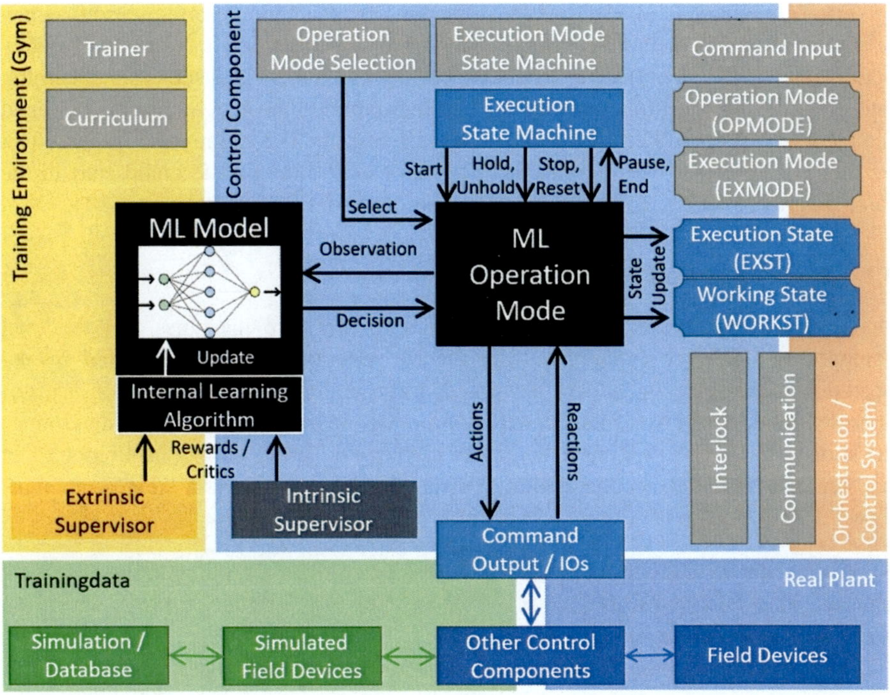

Fig. 4. Simplified interaction scheme for an ML operation mode in a control component.

The interaction of the ML operation mode with the command output determines the action space. As shown in Fig. 4 these **actions** are selected by the ML model, which are called **decisions** in this context. The second kind of interaction between the model and the operation mode are **observations** made by the CC. It is important to only use information available to the CC, in order to keep the component characteristic (handling as a single unit). In consequence, the **reactions** from the underlying CCs, mainly state changes, can be used. But, in general, any data that is available via the components' input signals (communication block) could be used. This might be necessary in order to include required information for the ML model, like the distance to the end position or the occupation states of additional conveyors in our scenario.

Another important input for the ML model, or rather the learning algorithm in particular, is an evaluation of the current environment state or the decisions taken by the ML model. Although these **rewards** or critics are only necessary in training phase, they can be very useful in the inference phase, e.g. as performance index. Usually the rewards are generated in the training environment by an extrinsic supervisor. To train the model in the production environment or with a hardware-in-the-loop (HiL) approach, it is best to only obtain values through the CC available information (intrinsic supervisor), analogous to the considerations for observations. In general, the states of the underlying CCs and interlock conditions are available and can be used for the intrinsic supervisor.

The ML model is also indirectly influenced by the trainer configuration, which selects the training methods and their parameters. In the scenario a proximal policy optimization (PPO) was chosen with default hyper parameters [18] such as the learning rate, etc. The task in learning may change during training time e.g., by reaching defined scores, which is controlled by the curriculum. Hence, the curriculum also has an effect on the supervisor and the training data. In our scenario the palette could start in the same position in every episode or in a random position, for example.

4.2 Mapping to Unified State Machines

Other interactions depicted above correspond to the state machines and the own components states (cf. Figure 4). After the ML operation mode is **selected** by the operation mode selection, it can be **started**. From then on to the completion, stop or abortion the control flow is handed over to the ML operation mode. So, it determines the **state update**. The working state may be used to explain the current behavior e.g., in our scenario it could set the output "current (conveyor) forward, next (conveyor) forward, shift1 up".

Execution States. The current execution state is determined according to the execution state machines, derived from PackML standard (see Fig. 1 right, c.f. [15]), where the states are interpreted as summarized in Table 1.

The main concept for the inner execution states is to use the suspending cycle to sync with the agent by requesting a new decision, especially if it is realized as an external model. This may lead to less time spent in the execute state, which has to be considered when building equipment efficiency measures and monitoring applications. Moreover, regular suspension may occur e.g., induced by interlock conditions. To separate those from decision requests, the working state may be used additionally e.g., showing "DECISION" while in suspending cycle.

Table 1. Interpretation of PackML execution states in control component ML operation modes.

State	Regular semantic (ISA-88)	Usage for ML operation mode in CC
Aborted, aborting, clearing	Aborting can be entered any time and will lead to a rapid safe stop. If the machine conditions allow for a reset the aborted state is entered and can be cleared	Usually, the outer states (c.f. Fig. 1) for aborting and stopping are handled by the CC frame, as a kind of safety mechanism. Hence, in the case of a stop or abort, the control flow will be withdrawn from the CC frame and the ML operation will be deselected. This will almost always lead to sending stop or abort commands to the underlying active units, as well as clearing and resetting them afterwards
Stopping, stopped, resetting	Stopping leads to a controlled stop, which can happen at any time except for the aforementioned aborting and corresponding states. A stopped machine can be brought to the idle state by requesting a reset command	
Idle	Idle indicates that the machine is ready to be started. It may perform basic tasks to maintain this state e.g., heating, spinning with idle speed	Idle is used to tell the CC what to do by selecting the operation mode. So, the control flow is handed to the operation mode selection. This can be used to build up the connection to the ML model or its runtime environment
Starting	Starting initiates the preparation steps to get the machine to the execute state	Starting may be used to initialize the ML model or the outputs, which corresponds to setting up the environment for inference or learning
Execute	The machine carries out its control and procedural tasks according to its mode. Actual material is processed	The ML operation mode carries out the actions for the ML model. It waits for one, specific or all command outputs to indicate that the underlying and occupied CCs are in an idle, completed or stopped execution state. Then the suspending state is initiated
Suspending	Suspending is initiated to let the machine pause for external process conditions. Machine speeds may be reduced to idle speed, etc. A typical condition is missing infeed	Suspending may be used to check if the goal is reached and to reassign the command outputs based on current conditions. It is a good point to add rewards for getting closer to the target. Then the suspended state is activated
Suspended	The machine waits for the external process conditions to return to normal and doesn't produce products	Suspended can be used to request a decision from the agent. It can also be used to reset and free CCs which are in stopped or completed state

(*continued*)

Table 1. (*continued*)

State	Regular semantic (ISA-88)	Usage for ML operation mode in CC
Unsuspending	Unsuspending is requested by the machine and prepares it for the execute state e.g., ramping up the machine speed	If the agent answers the decision request with an action request, the unsuspending state is initiated and the actions are converted to output commands by the macro. Execute state is reached if any underlying and occupied CC starts executing
Holding, held, unholding	The holding cycle is used to pause the execution because of a machine fault or external command to hold. The held state needs to be acknowledged from outside of the machine to get back to the execute state	If a hold is initiated, all active command outputs need to be set to hold in order to hold the current execution. The requests for decisions and consequently the requested actions of the ML model need to be interrupted. To let the agent start in proper conditions on unhold, the sending of observations should be carried on
Completing, complete	If the machine has completed a (self terminating) production process, it stops execution while completing and waits to be reset in completed state	The completing state is initiated if the target for the ML operation mode is reached. Hence, at least all information to check for the goal conditions need to be available for the internal supervisor, even in the inference phase

Execution Modes. Besides the execution states, the execution modes according to ISA-88 (c.f. [15]) may be used to unify the behavior of an ML powered CC in the following way:

- **Automatic** (Producing): Inference. The automatic mode is used for the inference phase, where the ML operation mode is used to control the real production.
- **Manual**: Heuristic. Manual mode is used to trigger possible actions of the ML operation mode from the command input or external operator interface. Hence, the action space has to be mapped to signal inputs or parameters of the ML operation mode. This is useful for testing the actions before training or after shifting from the gym or offline training, respectively, to the real plant e.g., after commissioning.
- **Semi-automatic** (Maintenance): The Semi-automatic mode may be used to test the trained model. For example, an acknowledge can be requested from the operator, before leaving the suspended state, after the agent has chosen an action. Thereby, the operator can check the decisions of the ML model step by step. Another use case is the online optimization via imitation learning. So, the operator picks or overwrites an action instead of the ML model in certain cases e.g., in difficult situation.

- **Simulation**: Learning. In the BaSys Project, another execution mode is proposed to show that the CC is using a simulation, redirecting its outputs or by simulating its behavior, without affecting the real plant. This may be used for online learning or proposing future action to the operator.

Additionally, an execution mode for learning or fine-tuning may be added as new designated mode, called "Learning" or "Optimization". In this execution mode, the real plant will be controlled by the ML model in training and further safety measures may be in place.

5 Summary and Outlook

In this contribution, we showed first results of designing and structuring control components for process control systems that are able to incorporate AI and ML algorithms and that are able to handle these functionalities in a structured and standardized way. Thereby, advanced data processing and control concepts from the area of AI and ML can be incorporated into industrial process control environments.

The considered technical example of coil transportation logistics may be simple, as the simulation of transportation effects can be done with most simulation physics' engines. Nevertheless, the provided concepts are independent and may be used with more sophisticated simulations, which may even be constructed using AI methods [20] e.g., system identification, which is a classical machine learning field. This raises the research question of how precise the models need to be for training purposes or whether ML models can be trained on general tasks and be fine-tuned later on the real plant, as it is already common practice in image classification methods.

Another open question is the direct deployment of trained machine learning models into state-of-the-art industry computers and PLCs. Here, virtualization and edge computing may be key solution approaches, as well as new dedicated PLC hardware.

As a next step, the provided concepts should be evaluated on more sophisticated simulations and in real application scenarios. This should include research on monitoring of the inference activity (i.e., the ML-based close-loop plant control) for safe operation and on explainable AI for a higher acceptance and incremental integration. Finally, state-of-the-art ML engineering workflows, data exchange formats, corresponding tools and courses need to be adapted to a partly automated learning process. The resulting solution would support empowering engineers and operators to focus on what the plant should do by parametrizing the targets rather than (re-)designing and maintaining the control application program.

Acknowledgement. The research leading to these results has been funded by the German Federal Ministry of Education and Research (BMBF) under grant agreement no. 01IS16022 (BaSys 4.0) and 01IS19022 (BaSys 4.2). More information can be found on the german project site: https://basys40.de.

References

1. Farshidian, F., Neunert M., Buchli, J.: Learning of closed-loop motion control. In: IEEE/RSJ International Conference on Intelligent Robots and Systems, pp. 1441–1446, Chicago IL, (2014). https://doi.org/10.1109/iros.2014.6942746
2. Thuruthel, T.G., et al.: Model-based reinforcement learning for closed-loop dynamic control of soft robotic manipulators. IEEE Trans. Robot. **35**(1), 124–134 (2019). https://doi.org/10.1109/TRO.2018.2878318
3. Vlachogiannis, J.G., Hatziargyriou, N.D.: Reinforcement learning for reactive power control. IEEE Trans. Power Syst. **19**(3), 1317–1325 (2004). https://doi.org/10.1109/TPWRS.2004.831259
4. Duan, J., et al.: Deep-reinforcement-learning-based autonomous voltage control for power grid operations. IEEE Trans. Power Syst. **35**(1), 814–817 (2020). https://doi.org/10.1109/TPWRS.2019.2941134
5. Matei, I., et al.: Deep learning for control: a non-reinforcement learning view. In: American Control Conference, pp. 2942–2948 (2020). https://doi.org/10.23919/acc45564.2020.9147287
6. Gros, S., Zanon, M.: Data-driven economic NMPC using reinforcement learning. IEEE Trans. Autom. Control **65**(2), 636–648 (2020). https://doi.org/10.1109/TAC.2019.2913768
7. Hafner, R., Riedmiller, M.: Reinforcement learning in feedback control. Mach. Learn. **84**(1–2), 137–169 (2011)
8. Fang, G., et al.: A reinforcement learning method to scheduling problem of steel production process. In: ISCME Journal of Physics: Conference Series, vol. 1486, p. 072035, (2020). https://doi.org/10.1088/1742-6596/1486/7/072035
9. Liu, C., Ding, J., Sun, J.: Reinforcement learning based decision-making of operational indices in process industry under changing environment. In: IEEE Transactions on Industrial Informatics (2020). https://doi.org/10.1109/tii.2020.3005207
10. Lingwei, Z., et al.: Scalable reinforcement learning for plant-wide control of vinyl acetate monomer process. In: Control Engineering Practice, vol. 97 (2020). https://doi.org/10.1016/j.conengprac.2020.104331
11. Iannino, V., et al.: A multi-agent approach for the self-optimization of steel production. In: International Journal of Simulation: Systems, Science & Technology (2018). https://doi.org/10.5013/ijssst.a.19.05.20
12. Schmidt, A., Schellroth, F., Riedel, O.: Control architecture for embedding reinforcement learning frameworks on industrial control hardware. In: Proceedings of the 3rd International Conference on Applications of Intelligent Systems (APPIS 2020). Association for Computing Machinery, Article 14, 1–6, New York (2020). https://doi.org/10.1145/3378184.3378198
13. Weichert, D., et al.: A review of machine learning for the optimization of production processes. Int. J. Adv. Manuf. Technol. **104**, 1889–1902 (2019). https://doi.org/10.1007/s00170-019-03988-5
14. Grüner, S., et al.: Cross-Industry State of the Art Analysis of Modular Automation. In: 21. Leitkongress der Mess- und Automatisierungstechnik, online (2020)
15. BaSyx CC. https://wiki.eclipse.org/BaSyx_/_Documentation_/_API_/_ControlComponent. Accessed 11 Dec 2020
16. Wagner, C., et al.: Uniform engineering for runtime environments. In: atp magazine, Bd. 61 Nr. 9 (2019)
17. CC Profiles Project. https://git.rwth-aachen.de/acplt/basys4.2/ccProfilesUA. Accessed 11 Dec 2020

18. Unity ML-Agents Code. https://github.com/Unity-Technologies/ml-agents. Accessed 11 Dec 2020
19. Prototype Source Code. https://git.rwth-aachen.de/acplt/basys4.2/PTSControl. Accessed 11 Dec 2020
20. Scheiderer, C., et al.: Simulation-as-a-service for reinforcement learning applications by example of heavy plate rolling processes. Procedia Manuf. **51**, 897–903 (2020). https://doi.org/10.1016/j.promfg.2020.10.126

Quality 4.0 - Transparent Product Quality Supervision in the Age of Industry 4.0

Jens Brandenburger[1(✉)], Christoph Schirm[2], Josef Melcher[2], Edgar Hancke[2], Marco Vannucci[3], Valentina Colla[3], Silvia Cateni[3], Rami Sellami[4], Sébastien Dupont[4], Annick Majchrowski[4], and Asier Arteaga[5]

[1] VDEH-Betriebsforschungsinstitut, Sohnstr. 65, 40237 Düsseldorf, Germany
jens.brandenburger@bfi.de
[2] thyssenkrupp Rasselstein GmbH, Koblenzer Str. 141, 56626 Andernach, Germany
{christoph.schirm, josef.melcher, edgar.hancke}@thyssenkrupp.com
[3] Scuola Superiore di Studi Universitari e di Perfezionamento Sant'Anna, Piazza Martiri della Liberta 33, 56127 Pisa, Italy
{marco.vannucci, valentina.colla, silvia.cateni}@santannapisa.it
[4] CETIC - Centre d'Excellence en Technologies de l'Information et de la Communication, Avenue Jean Mermoz 28, 6041 Charleroi, Belgium
{rami.sellami, sebastien.dupont, annick.majchrowski}@cetic.be
[5] Sidenor I+D, Barrio Urgate, 48970 Basauri, Spain
asier.arteaga@sidenor.com

Abstract. Progressive digitalization is changing the game of many industrial sectors. Focusing on product quality the main profitability driver of this so-called Industry 4.0 will be the horizontal integration of information over the complete supply chain. Therefore, the European RFCS project "Quality 4.0" aims in developing an adaptive platform, which releases decisions on product quality and provides tailored information of high reliability that can be individually exchanged with customers. In this context Machine Learning will be used to detect outliers in the quality data.

This paper discusses the intermediate project results and the concepts developed so far for this horizontal integration of quality information.

Keywords: Industry 4.0 · Quality supervision · M2M · Outlier detection

This project has received funding from the Research Fund for Coal and Steel of the European Union (EU) under grant agreement No 788552.

V. Colla and C. Pietrosanti (Eds.): ESTEP 2020, AISC 1338, pp. 54–66, 2021.
https://doi.org/10.1007/978-3-030-69367-1_5

1 Introduction

In a world where steel products can be acquired through platforms like Alibaba.com and the steel market is flooded with cheap steel from Chinese overcapacity, European Steel producers urgently need differentiation as distance or referencing are not a protection anymore [1].

From steel customers' point of view, one main reason for the decision for a specific supplier is trust in the fact that the delivered product fulfills his individual requirements. Consequently, only if the European Steel Industry succeeds to win customer-trust and solidifies client intimacy, it will achieve a durable competitive advantage and thus reduce pressure from world-wide imports. This need especially holds for markets like automotive industry were product and data traceability are gaining importance. But how to gain customer-trust and intimacy? Keys to success in this context are communication and information exchange, as the only feasible way to make steel customers rely on assured product quality is to share related quality information. On the other hand, customer-feedback can help the supplier as well to improve his own processes, so a bidirectional exchange of quality information would be a major advantage for both sides. In the context of recent discussions about Industry 4.0 this approach is referred to as the "horizontal integration over complete supply chains" and identifies one of the four main aspects characterizing this vision of digital transformation of manufacturing processes [2].

2 Problem Description

The digitalisation process often referred to as "Industry 4.0" has not fully reached the industrial practise in European Steel Industry yet. Although recently a great development of general-purpose IT-platforms and solutions for industrial data handling and treatment could be observed, in a recent study 62.6% of the steel customers assessed the digitalization level of the Steel Industry still as low [3].

This is because general-purpose IT-Platforms, such as, for instance, MindSphere from Siemens[1], Predix from General Electric[2], Ability™ Genix from ABB[3] and IBM Watson[4] or Open Source platforms like OpenIoT, Eclipse Kura or Kaa usually forget that a lot of work and process knowledge is mandatory to adapt existing products to complex industrial value-chains as formed by the steel manufacturing. As stated in [1], only if process knowledge is combined with IT expertise, value creation by means of digitalisation becomes possible. When thinking about the horizontal integration of quality information over the complete supply chain, it becomes obvious very quickly, that this idea implies much more than just the definition of some standard IT-interfaces and data formats for the exchange of quality data. At first, data characterizing the

1 https://www.siemens.com/global/en/home/company/topic-areas/digitalization/mindsphere.html.

2 https://www.ge.com/digital/predix.

3 https://new.abb.com/process-automation/genix.

4 https://www.ibm.com/watson/.

quality of a product is by far the most critical information gathered during the steel production process, especially with respect to the supplier-customer relationship. To exchange just this crucial kind of information, immediately raises the demand on information reliability to the utmost possible level. To share wrong quality information may cause severe customer uncertainty and long-term damaged customer confidence and should be avoided at all costs.

Consequently, a platform for the horizontal integration of quality information over the complete supply chain has to face three major challenges:

1. Guarantee reliability of quality information
2. Supervise, that the product and the quality information fulfil given specifications
3. Customize the quality data exchange to contain all relevant information for each individual order.

A solution realizing Quality 4.0 has to respect all these challenges in an integrated way to enable the evolution of efficient and quality-oriented future manufacturing.

3 The Quality 4.0 Approach

To realize the vision of horizontal integration of quality information through the complete supply chain, it is essential to guarantee quality data plausibility, to allocate products (coils, billets, bars) and data to customers/orders, and finally to compile and exchange the data. Therefore, the solution that will be developed within this project will consider all of these aspects, create methods for plausibility protection as well as for quality data management and compile the developed solutions into the Quality 4.0 platform shown in Fig. 1. Using a Service-Oriented-Architecture (SOA) the single components of the Quality 4.0 platform can be flexible combined and integrated in the existing IT infrastructure without relying on a single software vendor, product or technology.

Figure 1 shows the concept from the rolling point of view receiving slabs from a steel supplier and providing coils to a customer. The same concept can be applied to the steelmaking process for example. To respond to the considered needs of the project, the Quality 4.0 Platform will be composed with 3 service components (as illustrated in the Fig. 1 in framed bold red):

- Quality Data Generation Service (QGS) component generates the quality data and their plausibility values
- Quality Allocation Service (QAS) component allocates the customer order to a suitable product and selects the relevant associated quality data
- Quality Exchange Service (QXS) component exchanges the selected quality data compiled for each customer order.

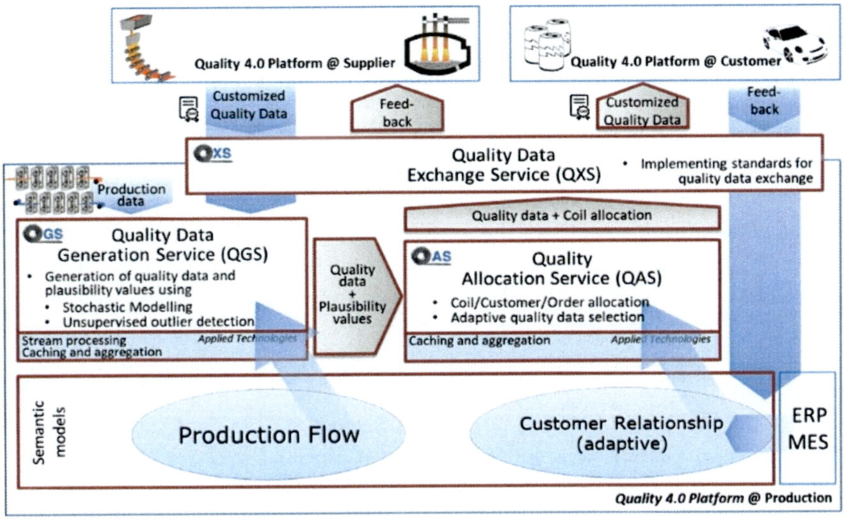

Fig. 1. Concept of Quality 4.0 architecture (Rolling example)

4 Quality Data Generation Service (QGS)

4.1 Data Plausibility Protection

In literature Commonly high-quality data is defined from the user's point of view as data that are "fit for use by data consumer" [4]. Reliability of data is seen as a "multidimensional, complex, morphing and goal-oriented concept" [5]. Consequently, the dimensions that have to be considered for data quality evaluation are specific for each data application and there is no common agreement in literature, which dimensions to choose in this definition. In [6] four dimensions, "Accuracy, Completeness, Timeliness, Consistency" have been discovered with major relevance in the context of industrial data quality assessment. This fact is taken into consideration by the papers dealing with the quality of industrial data. Examples from the oil and gas industry are given in [7] and [8], from the chemical industry in [9].

In the field of big data analytics, the consideration of the data quality is just at the beginning, as further challenges have to be met regarding data volume and diversity [10]. In this case increasing demands for the timeliness and the real-time processing of the data is foreseen, where solutions are still lacking.

4.2 Plausibility Value Calculation

The main functionality of the Quality 4.0-QGS is to estimate quality data from all available data sources and quantify the confidence of this estimation by means of a plausibility value (PV) to final guarantee reliability of quality information provided.

In the daily production aspects like accuracy, completeness, timeliness and consistency of quality related data have to be considered comprehensively for PV calculation. Particularly demanding in this context is the plausibility protection for data coming from automatic surface inspection systems (ASIS), as the quantification of ASIS accuracy is still a subject of ongoing research [11].

As a generic approach for the calculation of PVs the determination of plausibility values can be described as a function

$$PV := f(I) \rightarrow [0, 1] \tag{1}$$

In the context of this paper a value of 0 always means that the data is not plausible and a value of 1 stands for data which should be fully reliable. Depending on the investigated type of quality data, different input data domains may apply to I. Methods covering one dimensional input data ($I = \mathbb{R}$) can be applied directly to 1D quality data as well as indirectly to measurement position or area values originating from 2D quality measurements.

Table 1 shows a list of data plausibility measures used in the Quality 4.0 project. These methods are based on simple mathematical formulas and can be combined to a complex plausibility assessment (Fig. 2).

Table 1. List of data plausibility measures

	Function	Parameter
Constant	$f(x) := p_m \in [0,1]$	$p_m \in [0,1]$
Threshold-ing	$f(x, t_{min}, t_{max}) := \begin{cases} 0 & \text{for} \quad x < t_{min} \\ 1 & \text{for} \quad t_{min} \le x \le t_{max} \\ 0 & \text{for} \quad x > t_{max} \end{cases}$	$t_{min}, t_{max} \in I$
Fuzzy-based	$f(x, t_0, t_1, t_2, t_3) := \begin{cases} 0 & \text{for} \quad x \le t_0 \\ \frac{x - t_0}{t_1 - t_0} & \text{for} \quad t_0 < x < t_1 \\ 1 & \text{for} \quad t_1 < x \le t_2 \\ \frac{t_3 - x}{t_3 - t_2} & \text{for} \quad t_2 < x \le t_3 \\ 0 & \text{for} \quad x > t_3 \end{cases}$	$t_0, t_1, t_2, t_3 \in I$
Variation-based	$f(x) := \begin{cases} 0 & \text{for} \quad \max(W_x^n) - \min(W_x^n) = 0 \\ 1 & \text{otherwise} \end{cases}$	$n \in \mathbb{N}$
Data driven	$f(x) := PV(x)$	$PV: I \rightarrow [0,1]$

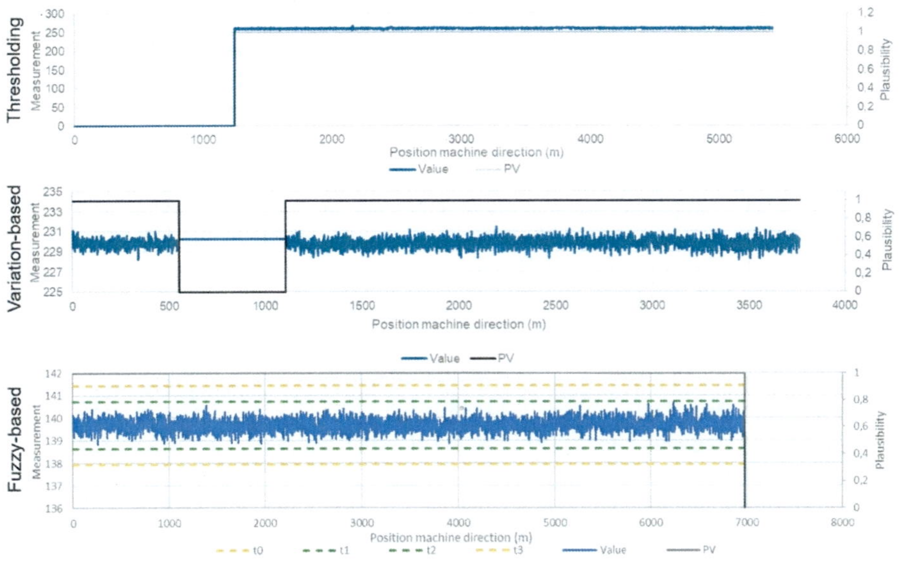

Fig. 2. Examples of position-based plausibility value calculation at a rolling mill

4.3 Outlier Detection

Quality related data collected throughout the production process are subjected to malfunctions in the sensors performing the measurement or other potential sources of errors in the measurements that lead to the creation of outliers. An informal definition of the concept of outlier involves its deviation from normality. There are multiple more formal definitions of outliers in literature that depend on the context [12] and can be grouped into 5 categories: distribution based, depth based, distance based, clustering based and density based. Each of these group focuses on a particular aspect of the deviation from normality.

In the context of the Quality 4.0 project an efficient and reliable detection of outlier quality measures plays a fundamental role as it avoids the sharing of misleading information about the sold products with customers that would degrade the customer-producer trust relationship. For this reason a module for the outlier-level calculation was embedded into the data plausibility component of the framework: this module is able to associate to each quality measure sample a outlier-risk value in the range [0; 1] in a completely automatic manner and thus realizes a data-driven PV calculation $f(x) = PV(x)$ as described in the previous section.

Due to the wide typology and variety of outliers there is not a commonly recognized approach for their detection that is efficient and reliable in any context. This fact is related to the above-mentioned different categories of definition that can be provided for the concept of outlier itself. Within the project the FUCOD[5] [13] algorithm for outliers detection was used. This method combines 4 existing approaches to outliers

[5] FUCOD = FUzzy Combination of Outlier Detection techniques.

detection with the aim of exploiting their strong points and avoiding their weaknesses according to processed data by means of a Fuzzy Inference System (FIS) that manages dynamically the contribution of each single method. FUCOD is designed to work with multidimensional data which means that the outlier-level will not be calculated only by considering the characteristics of the individual variables that form the quality data but also the interactions among them within the problem domain. These characteristics make the FUCOD approach particularly suitable to cope with industrial dataset as put into evidence by the numerous tasks, even within the steelmaking industry, where the method was successfully applied [14,15].

As shown in Fig. 3 FUCOD exploits 4 different methods belonging to different outlier detection algorithms families, namely:

- The Grubb's test, belonging to distribution-based methods, that takes into account the data deviation from a normal distribution
- A distance-based indicator that considers the distribution of the distances of data from their neighbors
- A clustering indicator that assesses the membership level of each data to its data-cluster within the dataset
- The Local Outlier Factor (LOF), a density-based approach that takes into account the density of data in the dataset domain.

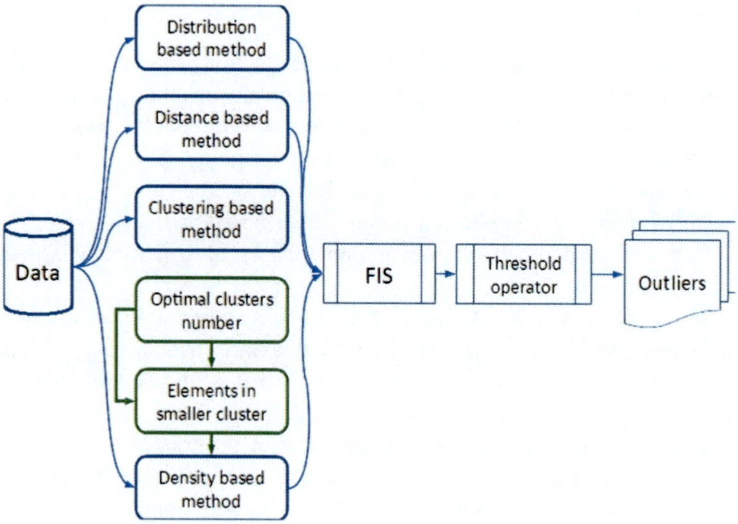

Fig. 3. Flowchart depicting the FUCOD algorithm for outliers detection

The output of single methods is suitably mixed by a FIS formed by a set of 6 rules and that returns, as final outcome, the outlier-level of each single observation in the dataset. In Quality 4.0 FUCOD is applied to the detection of outliers in data related to sensors measurements on manufactured coils. More in detail two different data sources were used in association to each coil: the first one consists of a set of 3 mono-dimensional measures (called P1, P2 and P3) concerning quality properties throughout the length of the coil, the second one (called P4) is a 2-dimensional map that covers the entire product surface. The FUCOD outliers detection was employed on a dataset of observation throughout the length of the coil where each data sample consists of P1, P2, P3 and the average of the P4 value along the coil width. The so-constituted 4-dimensional dataset is fed to FUCOD that automatically calculated the overall outlier level of each data sample corresponding to a set of measures over the coil length.

From a qualitative point of view the method achieves satisfactory results as it is able to point out data samples that seems to deviate from standard multi-dimensional distribution of observations in the dataset as depicted in Fig. 4. In general, on the tested coils, FUCOD detects a number of outlier samples that varies in the 0.01%–0.1% with respect to the number of observations within a dataset. The average computational time for each coil is 25 s that makes it acceptable to the integration in the Quality 4.0 framework.

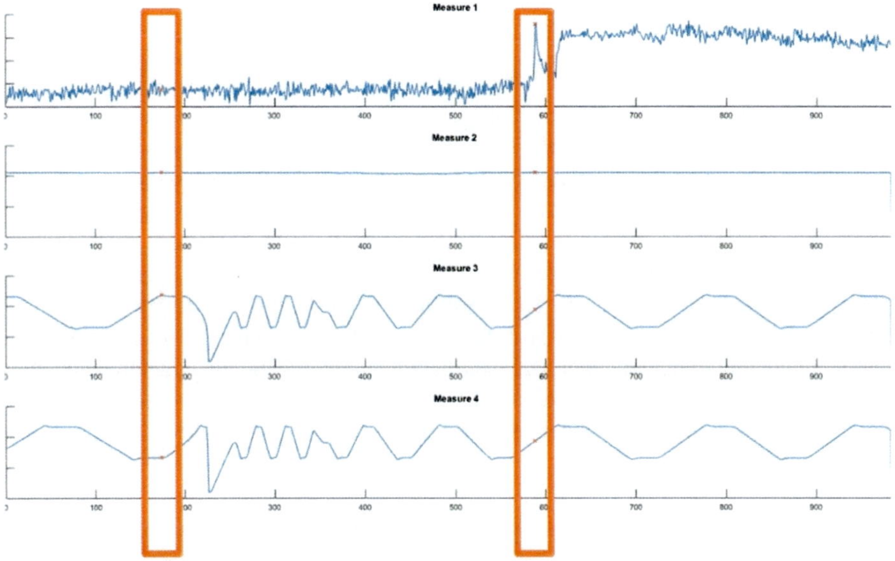

Fig. 4. Outcome of the FUCOD method on data belonging to a sample coil. Data samples with higher risk of being outliers are highlighted.

5 Quality Allocation Service (QAS)

The Quality 4.0-QAS realizes adaptive supervision of product quality by combination of quality data and knowledge about customer relations. As shown in Fig. 1 the Quality 4.0-QGS provides estimated quality data together with a dedicated plausibility value reflecting the confidence of the quality data. The Quality 4.0-QAS again receives this information and combines it with knowledge about the intended customer resp. Order. Based on this information the Quality 4.0-QAS has to decide autonomously if the product can be allocated to this customer resp. Order and selects the relevant quality data to exchange. Furthermore, in case of quality deficiencies the information provided by the Quality 4.0-QGS will be compiled by the Quality 4.0-QAS to create valuable feedback information for the supplier and send it to the QXS.

However, a system that intends to exchange relevant quality information between supplier and customer has to understand the meaning of "relevance". Thus, all information required for the determination of quality information relevance was semantically modelled based on available customer and order data. This customer relationship model implies to reflect the mutual trust between supplier and consumer and comprises the information as well under which conditions a coil fits to a certain order as which kind of quality information the Quality 4.0-QXS should provide to the specific customer. The level of customer intimacy was included in the model to be able to define the type, amount and plausibility of quality information that should be exchanged considering the interrelationship between producer and customer.

6 Quality Exchange Service (QXS)

Finally, based on the result provided by the Quality 4.0-QAS, the selected quality data will be compiled for each order individually and the data will be exchanged using a standard communication protocol. Therefore, the QXS is the only service that is accessible over plant boundaries and manages the quality data exchange between different instances of the Quality 4.0 platform. Thus, by exchanging quality and feedback information between supplier and customer, it realizes a fully bi-directional customer-oriented quality data exchange and establishes a common focus on product quality by horizontal integration. To determine a suitable IT standard for quality data exchange between the customers and the suppliers the following list of solutions has been defined and analyzed: the QDX[6], the STEP[7], and the Quality Tracking System[8]. As it turned out that no suitable free standard for quality data exchange exists, a Quality 4.0 specific IT standard will be defined and implemented in the Quality 4.0 framework.

7 Implementation

7.1 The FADI Framework

FADI is a customizable end-to-end big data platform enabling the deployment and the integration of open source tools in a portable and scalable way. It is a multi-tenant and multi-actors (i.e. business analyst, data scientist/engineer, IT admin, etc.) platform. The main features of FADI are five-fold (see Fig. 5): (1) collecting batch and stream data coming from various data sources, (2) storing data in different types of data stores, (3) processing data using ML and Artificial Intelligence (AI) techniques, (4) visualizing and analyzing data in a user Web interface, and (5) generate and publishing reports.

[6] VDA-QDX: https://portal3.gefeg.com/vdaqdx/page/home.

[7] https://www.steptools.com/stds/step/.

[8] https://www.eurofer.eu/Issues%26Positions/Quality%20Tracking/Quality%20Tracking.fhtml.

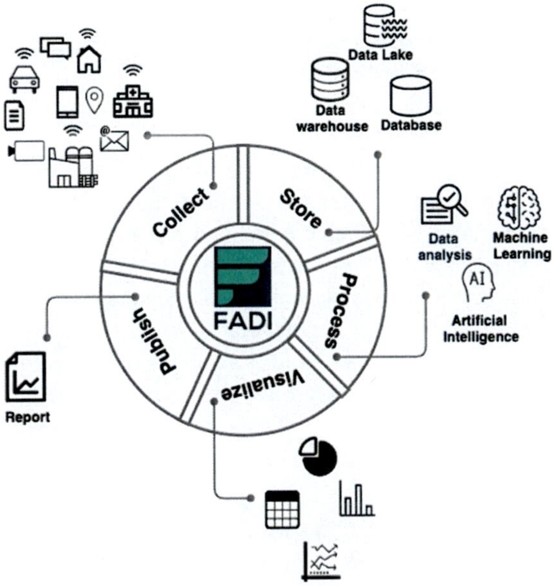

Fig. 5. An overview of FADI's features

7.2 A Sample Implementation of FADI

Some of the innovative aspects of FADI is the customizability, modularity and extendibility. Indeed, it provides a flexible software architecture that enables to easily integrate various open source tools dedicated to the management of Big Data and third-party tools (i.e. QGS, QAS, QXS, and Modoboa). Indeed, we present the implementation of FADI that allows to set-up the Quality 4.0 platform. In the following, we introduce how the FADI platform has been customized in order to put in place the Quality 4.0 and support the requirements of the project (see Fig. 6):

- For the **internal administration services**, we propose to use the following tools: The ELK[9] tools suite is used to manage the logs. ELK stands for the three open source tools: ElasticSearch to search information in the collected logs, Logstash to transport and process logs by collaborating with the Elasticsearch tool, and Kibana to visualize the logs and the results of its analysis. Then, Zabbix[10] is used to set-up the monitoring in the FADI infrastructure and OpenLDAP[11] is used to provide Identity and Access Management of the different users of the platform. Finally, we propose to configure the server mail Modoboa[12] in order to enable the exchange of the Quality 4.0 certificate and the feedback between the customer and the producer via the QXS.

[9] https://www.elastic.co/what-is/elk-stack.

[10] https://www.zabbix.com/.

[11] https://www.openldap.org.

[12] https://modoboa.readthedocs.io/en/latest/.

- We propose to use the Apache Nifi[13] tool to set up the **data integration** layer. It automates the data flow between systems: both with external systems (e.g. sensors data sources) and between internal systems (e.g. the various analysis and storage components). Apache Nifi is connected to the QGS in order to compute the PVs, to detect the outliers and to store raw and computed data.
- For the **data analysis** layer, we propose to use and integrate in the FADI platform the QGS and the QAS components. The QGS computes the different metrics. Whereas the QAS evaluates the quality of a given product and prepares the Quality 4.0 certificate.
- For the **data storage** layer, we propose to use the PostgreSQL[14] which is an object-relational database management system. It has been chosen because it contains advanced database features (e.g. aggregation, parallel queries, replication system, etc. This database will be used in order to store all the data (e.g. the production flow, customer relationship, ERP/MES, etc.) and the semantic models.

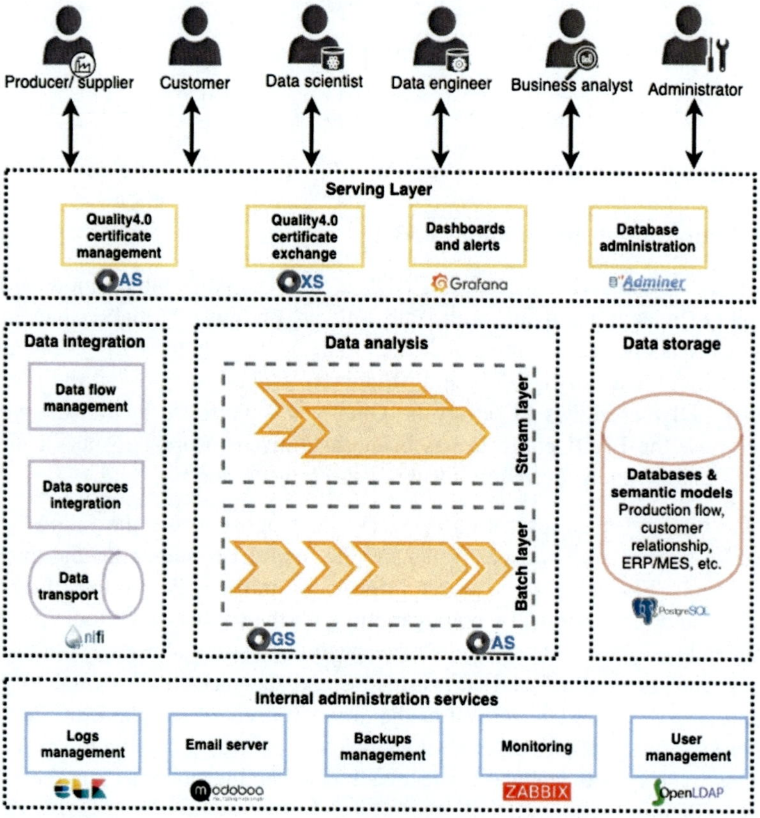

Fig. 6. Quality 4.0 platform implemented using the FADI platform

13 https://nifi.apache.org/.

14 https://www.postgresql.org/.

- For the **serving layer**, we propose to use Adminer[15] to manage the PostgreSQL database in a web interface. Grafana[16] is used to explore and visualize data. Then, the QAS interface is available to control the data quality allocation. Finally, the QXS interface is used to send and receive the Quality 4.0 certificate and the customer feedbacks.

7.3 The CI/CD Pipeline of FADI

To ensure speed and quality of the system's lifecycle, the Quality 4.0 platform relies on the DevOps approach that combines development and operations processes: plan, code, build, test, release, deploy, operate, monitor.

Those processes are automated through Continuous Integration and Delivery (CI/CD) pipelines to facilitate frequent iterations on integration between the platform components (e.g. QAS, QGS, QXS, data sources, data sinks, etc.) and the deployment of the solution to various environments (e.g. testbeds, servers in factories, etc.). Figure 7 illustrates the services and technologies that support the DevOps processes of the platform:

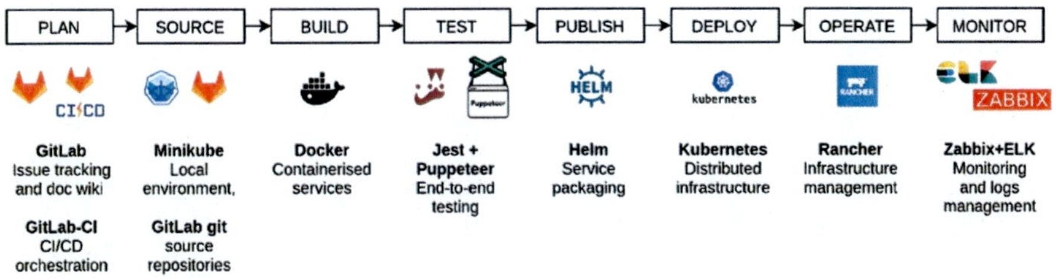

Fig. 7. Quality 4.0 DevOps CI/CD processes

8 Conclusion

Only if the European Steel Industry succeeds to win customer-trust and solidifies client intimacy, it will achieve a durable competitive advantage and thus reduce pressure from world-wide imports. Furthermore, it is of strategic importance for the European steel industry to proactively promote such a common platform instead of reacting on specific customer demands. In the RFCS project Quality 4.0 presented in this paper an adaptive platform tailored to the demands of the steel industry will be developed that realises the horizontal integration of quality information across the entire supply chain. In detail this new platform will provide quality information of high reliability, improved decisions on the reached product quality and automatic data exchange adapted to existing customer and order information.

[15] https://www.adminer.org/.

[16] https://grafana.com/.

References

1. Bonnaud, E., De Thieulloy, G., Lecat, A., Saint-Aubyn, J., Charpentier, A.: THINK ACT - Weathering the Steel Crisis. Roland Berger, Paris (2016)
2. acatech – Deutsche Akademie der Technikwissenschaften e.V., Umsetzungsempfehlungen für das Zukunftsprojekt Industrie 4.0, Frankfurt Main (2013)
3. IW Consult GmbH: Potentiale des digitalen Wertschöpfungsnetzes Stahl, WV Stahl (2017)
4. Wang, R.Y., Strong, D.M.: Beyond accuracy: what data quality means to data consumers. J. Manage. Inf. Syst. **12**(4), 5–33 (1996)
5. Dasu, T., Johnson, T.: Exploratory Data Mining and Data Cleaning. Wiley, Hoboken (2003)
6. Gertz, M., Tamer Ozsu, M., Saake, G., Sattler, K.: Data Quality on the Web, Report on the Dagstuhl Seminar (2004)
7. Schlumberger, R.R.: Facilitating data quality improvement in the oil and gas sector. In: SPE Asia Pacific Oil and Gas Conference and Exhibition, Perth, Australia (2008)
8. Hubauer,T., Lamparter, S., Roshchin, M., Solomakhina, N., Watson, S.: Analysis of data quality issues in real-world industrial data. In: Conference of the Prognostics and Health Management Society, New Orleans, USA (2013)
9. Jiang, T.W., Stuart, P.R., Chen, B., Jasim, K.: Strategy for improving data quality for a kraft pulp mill recausticizing plant. In: Proceedings of the FOCAPO Conference, Coral Springs Florida (2003)
10. Cai, L., Zhu, Y.: The challenges of data quality and data quality assessment in the big data era. Data Sci. J. (2015)
11. Brandenburger, J., Krambeer, H., Schirm, C., Jonker, K., Mißmahl, W., Nörtersheuser, M., Kogler, A., Ebner, A.: Towards measuring surface quality by means of automatic surface inspection systems. In: 4th European Steel Technology and Application Days, Düsseldorf (2019)
12. Barnett, V., Lewis, T.: *Outliers in Statistical Data*, Wiley Series in Probability and Mathematical Statistics. Wiley, Chichester (1994)
13. Cateni, S., Colla, V., Vannucci, M.: A fuzzy logic-based method for outliers detection. In: Artificial Intelligence and Applications, pp. 605–610, February 2007
14. Dimatteo, A., Vannucci, M., Colla, V.: Prediction of hot deformation resistance during processing of microalloyed steels in plate rolling process. Int. J. Adv. Manuf. Technol. **66**(9–12), 1511–1521 (2013)
15. Colla, V., Bioli, G., Vannucci, M.: Model parameters optimisation for an industrial application: a comparison between traditional approaches and genetic algorithms. In: Proceedings - EMS 2008, European Modelling Symposium, art. no. 4625243, pp. 34–39 (2008)

19. Vannucci, M., Colla, V.: Fuzzy adaptation of crossover and mutation rates in genetic algorithms based on population performance. J. Intell. Fuzzy Syst. **28**(4), 1805–1818 (2015)
20. Colla, V., Valentini, R., Vannucci, M.: A model for predicting residual hydrogen content in blooms and billets stacked in large industrial piles. Steel Res. Int. **89**(10), 1800155 (2018)
21. Vannocci, M., Ritacco, A., Castellano, A., Galli, F., Vannucci, M., Iannino, V., Colla, V.: Flatness defect detection and classification in hot rolled steel strips using convolutional neural networks. In: Lecture Notes in Computer Science. LNCS, vol. 11507, pp. 220–234 (2019)
22. Cateni, S., Colla, V., Vannucci, M., Vannocci, M.: Prediction of steel hardenability and related reliability through neural networks. In: Proceedings of the IASTED International Conference on Artificial Intelligence and Applications, AIA 2013, pp. 169–174 (2013)

The Use of Advanced Data Analytics to Monitor Process-Induced Changes to the Microstructure and Mechanical Properties in Flat Steel Strip

Frenk Van Den Berg[1(✉)], Danique Fintelman[1], Haibing Yang[1], Claudio Mocci[2], Marco Vannucci[2], and Valentina Colla[2]

[1] Tata Steel, Ijmuiden, The Netherlands
frenk.van-den-berg@tatasteeleurope.com
[2] Scuola Superiore Sant'Anna, Pisa, Italy
{claudio.mocci,marco.vannucci,
valentina.colla}@santannapisa.it

Abstract. In the production of cars, steel is the preferred material for the body because of its high strength, good formability and excellent recycling opportunities. In order to have a smooth production when pressing automotive body parts, the mechanical forming properties of the steel strip need to be uniform and consistent. These properties are governed by the microstructure of the steel. The final microstructure is largely determined by both the thermal process conditions during the continuous annealing process and the incoming state of the material. This paper describes the attempt to monitor the incoming state using the IMPOC device prior to the annealing process. Using advanced analytics methods, these data have been searched for correlations with process data and data from a second IMPOC device at the end of the production line.

Keywords: Continuous annealing · Inline monitoring · Variable selection · Data mining · Data analysis · Property prediction

1 Introduction

The steel industry is facing increasingly stringent requirements on the quality of their products. In particular for strip produced for the automotive industry, the process windows need to be accurately controlled to meet the tight customer demands on consistency and uniformity in mechanical properties, which are critically determined by the chemistry and processing parameters during the manufacturing chain. To support the quality management, the microstructure and mechanical properties are monitored inline (i.e. in real-time during production) by means of non-destructive measurement techniques. The development and use of these systems have been the subject of a number of collaborative projects in the RFCS[1] framework, like [1–5].

[1] RFCS = European Research Fund for Coal and Steel.

V. Colla and C. Pietrosanti (Eds.): ESTEP 2020, AISC 1338, pp. 78–91, 2021.
https://doi.org/10.1007/978-3-030-69367-1_7

One of the systems which has made it to commercialization is the IMPOC system [6, 7]. Since data from the IMPOC system is used in the advanced analytics in this paper, its operation principle is described concisely below. Further details can be obtained via the company EMG [8] which markets the device. The IMPOC system consists of 2 identical subsystems located at the top and bottom sides of the strip at a typical working distance of 25 mm. Each subsystem is composed of an excitation coil and, approximately 30 cm downstream (in strip movement direction), a sensing part. The excitation coil produces a short magnetic pulse causing the ferromagnetic steel to locally and temporary reach magnetic saturation induction. The excitation pulse quickly extinguishes and the magnetic field above the strip is then governed by the remnant part of the magnetization spot inside the steel strip. The latter magnetic field is picked up by the sensing part of the system. Since the strip moves with a typical speed of 3 m/s through the IMPOC system, the sensor will record a magnetic pulse. The maximum of this magnetic pulse is called the IMPOC upper (lower) value, where upper (lower) refers to the upper (lower) subsystem. To minimize effects from changes in the pass line, the average is taken of the IMPOC upper and IMPOC lower values, and this final value is then called the IMPOC value. Modelling of the IMPOC response has been the subject of papers by CEA [9, 10].

Empirical studies have shown that the IMPOC value and the tensile strength of steels are strongly correlated [11–13]. This relationship has triggered studies – some involving advanced analytics like multiparameter regression analysis, Self-Organizing Maps (SOM) and neural networks – to relate IMPOC values with process parameters or with other mechanical properties, as well as studies investigating the uniformity of coil [5, 11, 13–16].

The present paper also describes a study relating process parameters with the inline measured IMPOC value. The novelty of this work is that the dataset not only contains IMPOC data recorded at the exit of the line (which is the case with all other studies) but also IMPOC data at the line entry. The IMPOC system at the entry (IMPOC_ENTRY) thus gives a fingerprint of the microstructural state of the incoming material, and the IMPOC system at the exit (IMPOC_EXIT) of the microstructural state of the material after processing in the HDG. As far as we know, this is the first time that a study of such a product quality measurement at both entry and exit of a single production line is reported. A potential business application could be to adjust the process based on the monitored incoming material state in order to obtain more uniform properties of the final product, either within a single coil, or within coils of a given grade, e.g. by the use of adaptive control algorithms.

In this work, in the context of the application of advanced analytics techniques within the steel industry outlined in [17, 18], different approaches to variables selection and modelling will be used for a twofold scope. On one hand they put into evidence the relation among the huge amount of measures that are collected throughout the processes which are hard to discover even by experts due to their dimensionality and volume. On the other hand, the modelling techniques can help users in monitoring and control tasks by means of the application of advanced analytics or soft sensors as for instance in [19, 20].

The paper is divided as follows: Sect. 2 describes the necessary steps in data cleaning and pre-processing; next, Sect. 3 highlights the three approaches followed to select the input variables for the prediction models, i.e. process-knowledge based selection and two data-driven approaches, namely a Genetic Algorithm based method and a Decision Tree based method. Then, in Sect. 4, results are presented from several different prediction models tested in combination with the 3 different sets of input variables. The tested prediction models are regression models of polynomials of the 1st, 2nd and 3rd order and a 2-layer feed-forward neutral network. Finally, the results are discussed and concluded in Sect. 5.

2 Data Preparation and Pre-processing

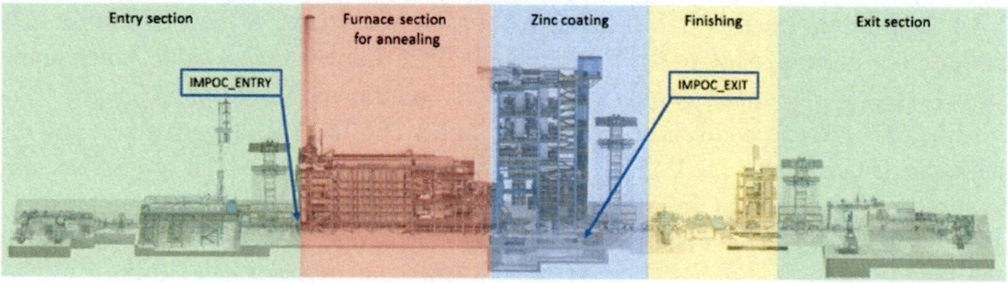

Fig. 1. Global layout schematic of a HDG plant, with the different sections indicated. Also indicated are the positions of the two IMPOC systems installed. See text for further details.

This work is based on data collected by Tata Steel Europe during the standard operation of the Hot-Dip Galvanizing (HDG) plant, of which an illustration is depicted in Fig. 1. Main functions of a HDG line are the annealing of the steel strip to regain the forming properties after cold-rolling, the coating of the strip with zinc for corrosion protection and, in the finishing section, the levelling to remove the yield-point elongation and obtain flat strip. The final product is shipped as coils to customers in the automotive and building sectors. The position of the IMPOC_ENTRY is just prior to the furnace section, whereas the IMPOC_EXIT is located after zinc coating and cooling down of the strip, but prior to the finishing section, see also Fig. 1.

In this HDG plant, process data and product data are collected on a regular basis. For the current analysis, a data set has been selected which contains about 75000 observations related to 81 coils of a single grade of high-strength-low alloy steel. The sampling rate is 2 m, and all data are synchronized.

A complete list of the process parameters subjected to analysis is shown in Table 1 together with their short description.

Before starting with the variable selection and modelling tasks that are the core of this work, the data have been pre-processed in line with Tata Steel expert suggestions in order to limit the presence of noise, errors and unreliable samples in the data that are proven to negatively affect subsequent data-driven approaches [21].

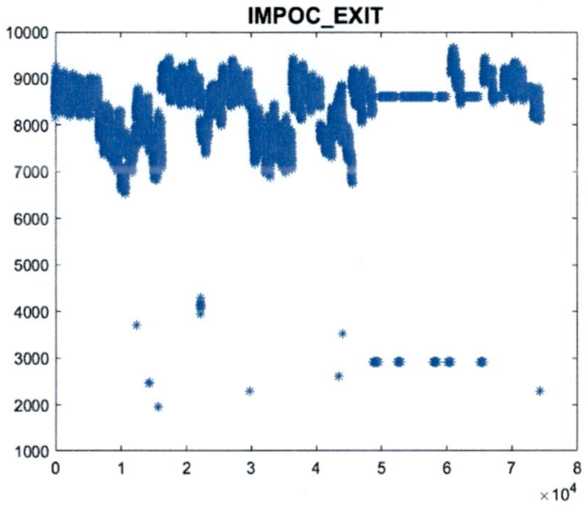

Fig. 2. IMPOC_exit before pre-processing.

The pre-processing steps can be summarized as follows:

- All samples referred to the first and last 100 m of each coil have been cut-off from the dataset because of their unreliability due to the presence of weldings.
- Samples characterized by a value of IMPOC_EXIT (shown in Fig. 2) lower than the nominal threshold of 5000 have been removed.
- Removal of data which had been interpolated by the recording system when data communications were interrupted. This phenomenon is also illustrated in Fig. 2 by the stationary behavior of the data.

The resulting dataset, after being pre-processed, contains about 40000 observations that are related to 44 coils.

3 Selection of Input Variables

This work focuses on the modelling of the value of the IMPOC signal at the exit of the HDG by exploiting all the available process parameters from the preceding process as input variables for an arbitrary model. In this context the collection of the plant process data offers a multitude of variables of which only a part is actually related to the *IMPOC_EXIT* and eventually some of these variables are redundant being correlated with each other. Since it was demonstrated that in data driven applications the use of *unnecessary* variables as model input is detrimental [22], a variables selection step was performed prior to the models tuning phase. Three different approaches described in the following paragraphs have been tested for this purpose:

- Process knowledge based (Sect. 3.1)
- Genetic algorithm (Sect. 3.2)
- Decision tree (Sect. 3.3)

followed by a final filtering step aiming to the elimination of eventually redundant variables among the selected ones described in (Sect. 3.4).

3.1 Process Knowledge

A first variable selection has been done directly by the process experts in Tata Steel. A list of the estimated most influential input variables is shown in Table 1 in the column denoted by PK (Process Knowledge).

3.2 Genetic Algorithm

This approach exploits the Give-a-Gap algorithm [23] that is based on the use of Genetic Algorithms (GAs) and faces the problem of variable selection as an optimization problem. The optimal solution is represented by the subset of variables - among the whole list of candidates - that minimizes the prediction error obtained by an arbitrary data driven model (in this work a 2nd order polynomial) that acts as the fitness function of the GA. This approach, whose flowchart is depicted in Fig. 3, achieved interesting results when coping with literature datasets and when applied to industrial problems [24–26].

The Give-a-Gap codes a generic solution as a string of bits where each bit corresponds to a candidate variable and where 1-valued bits represent the variables actually selected according to the solution. The GA optimization engine explores in an efficient manner the domain of the solution (which includes 2n potential candidates) by suitably mixing through the GA crossover operator the more efficient ones (those with a high value of fitness) and explores possibly promising sets of variables by mutating the existing ones. After a predetermined number of generations, the GA stops and returns the optimal solution. In this work an ad hoc fitness function for the assessment of candidate solutions was employed. Such function, given the candidate set *V*, trains a polynomial model by using only *V* and then measures the performance *e* of the so-obtained model then, in order to assign higher fitness to solutions that achieve a lower error, the *1/e* value is returned as fitness of the candidate set *V*.

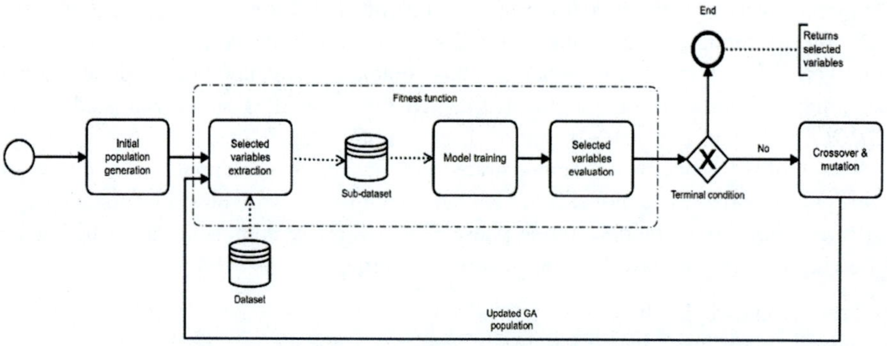

Fig. 3. Genetic algorithm for variable selection.

3.3 Decision Trees

In this work Decision Trees (DTs) were used for variable selection purposes as well. This approach gives a clear insight of the type of relation that binds the input variables to the target allowing a greater comprehension of both the IMPOC signal evolution during the HDG process and of the results of the selection step.

This experimental method uses the whole training dataset to create a DT for the *IMPOC_EXIT* prediction by means of the well-known C4.5 algorithm. The resulting DT is formed by a large number of nodes. Each node represents a *split* in the rationale for the determination of the target value that takes into account an input variable within an *if-then* rule. The nodes, if browsed from the root to the leaves of the tree, define the input-target relation through a set of readable rules.

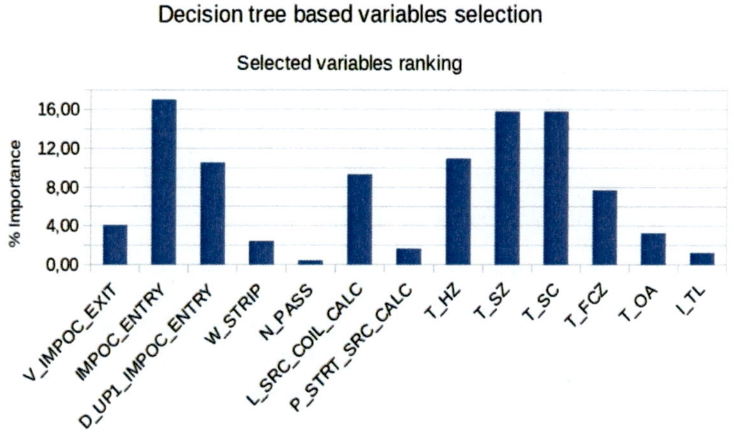

Fig. 4. Relative importance of variables selected through the use of DT.

An intuitive measure of the importance of the potential input variables for the modelling of the *IMPOC_EXIT* signal can be obtained by counting the number of appearances of each single variable within the nodes of the tree. This measure is represented in Fig. 4 where the relative importance of selected variables is expressed as a percentage.

3.4 Variables Redundancy Filter

At the end of the GA and DT based variable selection processes it is possible that the set of the selected variables is over-dimensioned containing some variables highly related amongst each other. This situation, although non detrimental from a point of view of subsequent model performance, can unnecessarily extend the training time. In order to avoid this drawback a final pre-processing step for the variable selection process is to eliminate the redundant variables from the set of the selected ones by keeping a minimal subset of independent variables. This filtering process is mainly composed by two subsequent steps: the first one envisages the detection of the *clusters* of highly interrelated variables, the second step, extracts from these clusters the minimum subset of variables to represent the cluster and that will be added to the final set to feed the predictive models.

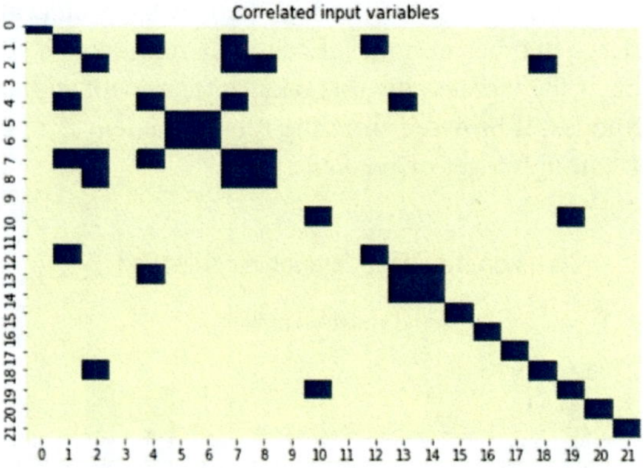

Fig. 5. Matricial representation of the redundancy among the input variables selected by the DT variables selection approach.

Clusters of correlated variables are found by calculating the correlation among the variables and by applying a suitable threshold: in this work the *Spearman* correlation coefficient was used and two variables are considered correlated if the correlation coefficient is higher than 0.95 and if its significance (also known as p-value) is lower than 0.05. These strict conditions lead to the determination of a *network* of related variables. An exemplar graphical representation of the correlation matrix once thresholds are applied is shown by the binary plot in Fig. 5 for the case of decision trees. In the figure the redundancy between two variables – according to the adopted thresholds - is represented by the blue squares.

In order to retain a minimal yet representative set of variables from these networks a *dominating set* approach [27] to the problem was used that led to the selection, within each network, of the minimum cardinality subset of variables that are related to all the other variables in the network.

Table 1 summarizes the result linked to Fig. 5 for the GA dataset, as well as the input variables selected by the decision tree (DT).

Table 1. Overview table with the process and product parameters in the dataset, with their description. The three columns at the right show which variables have been selected as input variables (**I**) and target values (**T**) by the different selection methods.

	Parameter name	Description [unit]	Selected input variables, by method		
			PK	GA	DT
IMPOC_EXIT parameters	IMPOC_EXIT	IMPOC – EXIT value (average of upper and lower sensors)	T	T	T
	MR_IMPOC_EXIT	Setting for IMPOC_EXIT			
	V_IMPOC_EXIT	Line speed at location of IMPOC EXIT [m/min]		I	I
	D_UP1_IMPOC_EXIT	IMPOC - EXIT: Distance between strip and IMPOC upper sensor (left side) [mm]			
	D_UP2_IMPOC_EXIT	IMPOC - EXIT: Distance between strip and IMPOC upper sensor (right side) [mm]			
IMPOC_ENTRY parameters	IMPOC_ENTRY	IMPOC – ENTRY value (average of upper and lower sensors)	I	I	I
	V_IMPOC_ENTRY	IMPOC - ENTRY: line speed at position of IMPOC ENTRY [m/min]		I	
	D_UP1_IMPOC_ENTRY	IMPOC - ENTRY: Distance between strip and IMPOC upper sensor (left side) [mm]			I
	D_UP2_IMPOC_ENTRY	IMPOC - ENTRY: Distance between strip and IMPOC upper sensor (right side) [mm]		I	
Strip Dimension	L_STRIP	Length of strip [m]			
	W_STRIP	Strip width [* 0.1 mm]			I
	T_STRIP	Thickness of strip [um]			
Pre-HDG parameters	N_PASS	Number of re-coiling steps prior to HDG			I
	L_SRC_COIL_CALC	Calculated length of source coil, based on weight / (density * (length * width)) [m]		I	I

	P_STRT_SRC_CALC	Calculated position of the source (i.e. the mother coil) in the input material (the daughter coil) [m].			I
Annealing Temperatures	T_HZ	Temperature at heating zone [°C, scaled]	I	I	I
	T_SZ	Temperature at soaking zone [°C, scaled]	I	I	I
	T_SC	Temperature at slow cooling zone [°C, scaled]	I	I	I
	T_FCZ	Temperature at fast cooling zone [°C, scaled]	I		I
	T_OA	Temperature at overaging zone [°C, scaled]	I		I
Post-HDG parameters	I_TL	Index of tension levelling of HDG coil			I
	RED_PERC_TL	Reduction percentage tension lev-eller [%]		I	
	RED_PERC_TM	Reduction percentage temper mill [%]			

4 Prediction Models and Results

Prediction models are based on polynomials (first, second and third orders) and Feed-Forward Neural Networks. All datasets have been divided into three subsets for training, test and validation procedures as follows:

- Training set: 60%
- Validation set: 20%
- Test set: 20%

For each dataset, the same training set has been exploited to train the models. The validation to validate them and the test set to see how models respond to unseen input data samples.

The goodness of all the predictions is evaluated through the error measure reported in Eq. (1) where t is the IMPOC_EXIT target vector, $\hat{t}$ is its model prediction, n is the overall number of observations and σ is the operator for the calculation of target standard deviation. Equation (1) puts into relation the discrepancy between measured and predicted *IMPOC_EXIT* values with its actual standard deviation in order to highlight the improvement in the target estimation due to the model use:

$$\frac{\sum_{i=1}^{n}|\hat{t}_i - t_i|}{n} \cdot \frac{1}{\sigma(t)} \tag{1}$$

Four models have been trained for each dataset for the prediction of the *IMPOC_EXIT*. Three are polynomials (first, second and third order) and one is a two layers Feed Forward Neural Network FFNN with 10 neurons in its hidden layer.

The results obtained with the three datasets given by variable selection algorithms shown in Sect. 3 are compared to those obtained using the dataset composed of all process parameters. Moreover, another comparison has been done excluding from each input dataset the parameter *IMPOC_ENTRY* in order to study the effect of the *IMPOC_ENTRY* on the goodness of the prediction.

Table 2 shows the errors for the predictions by various combinations of the applied models with the input variable sets. The top rows (red & black text color) show the errors for training & test sets related to datasets shown in Sect. 3, whereas the bottom rows show the errors on the predictions (purple/blue for training/test sets) obtained excluding *IMPOC ENTRY* from each input dataset.

Table 2. Errors for the different models (rows) and different input variable sets (columns): top (red & black) versus bottom (purple & blue) rows: inclusion versus exclusion of IMPOC_ENTRY as input parameter. Left (red & purple) versus right (black & blue) columns: Training Set versus Test Set.

Errors		**PK dataset 6 / 5 input vars**		**GA dataset 9 / 8 input vars**		**DT dataset 13 / 12 input vars**		**Whole dataset 22 / 21 input vars**	
		Training	*Test*	*Training*	*Test*	*Training*	*Test*	*Training*	*Test*
Polynomial 1st order	*Incl*	0.490	0.490	0.369	0.369	0.330	0.328	0.302	0.303
	Excl	0.576	0.569	0.399	0.400	0.359	0.354	0.325	0.324
Polynomial 2nd order	*Incl*	0.465	0.471	0.310	0.312	0.290	0.292	0.259	0.256
	Excl	0.537	0.538	0.352	0.350	0.331	0.333	0.289	0.288
Polynomial 3rd order	*Incl*	0.464	0.466	0.310	0.312	0.280	0.289	0.253	0.254
	Excl	0.532	0.534	0.349	0.350	0.330	0.331	0.286	0.287
FFNN	*Incl*	0.232	0.231	0.160	0.166	0.145	0.148	0.165	0.173
	Excl	0.306	0.303	0.188	0.192	0.180	0.180	0.189	0.189

Figure 6 shows the predictions (*y-axis*) and measured (*x-axis*) values of the best model for each dataset. The best result has been obtained using as input those variables selected by the decision tree algorithm.

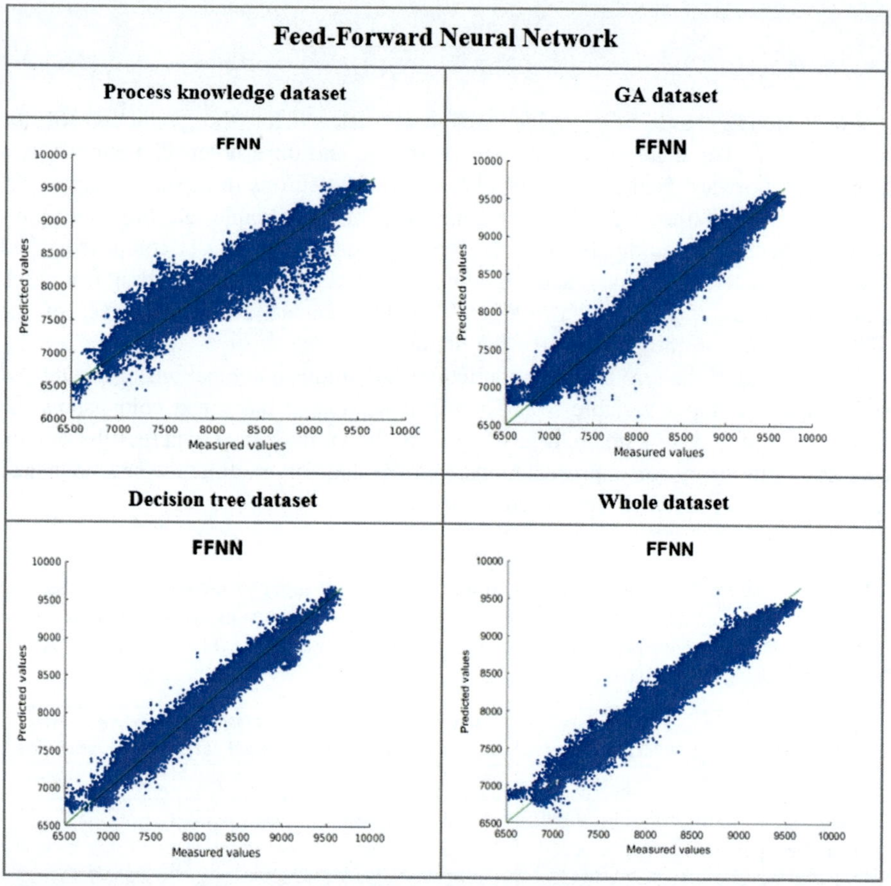

Fig. 6. Predicted (by FFNN model) versus measured IMPOC_EXIT, using the 4 different input variable sets.

5 Discussion and Conclusions

When considering the results of Table 2, a comparison between the left and right numbers in a given cell appear to be very close. The errors on the training and test sets are thus almost equal, indicating that – in all cases – the models have not been overfitted.

As discussed, polynomial models have been investigated of order 1, 2 and 3. Whereas the 2nd order polynomial models result in lower errors than the 1st order polynomial models, increasing the order further to the 3rd order hardly give any further improvement. As expected, lower errors are always obtained with FFNN because it is more suitable to model non-linear behaviors. Still, use of any kind of these prediction models for data being outside the trained input parameter ranges can result in unstable results, hence their value in process control is limited to commonly used process windows only. However, in the practice of steel strip processing, where many different grades are produced on a single production line in fairly arbitrary sequences, process windows tend to vary continuously. Although the FFNN model performs best in terms of the obtained error, a disadvantage is that, being a black box model, it cannot be directly inverted to a real-time model for process control. It would require a meta-model, in which the input parameter space is explored (within the validity window) to understand the relative importance of each input on the target results. On the contrary, the (2nd order) polynomial model offers the possibility to study its partial derivatives, and hence the effect of the input parameters on the target, in a direct manner. Such a knowledge could be readily transferred into process control models, with for example the objective to keep the target value as constant as possible within other process constraints.

When comparing the different input variable sets, we can observe that for the polynomial models, as expected, the higher the number of input variables, the lower the error. Still the GA dataset, having the lowest number (9) of input variables amongst the data-driven selection methods, performs only ~10% worse (in terms of the error) as the DT selected dataset with 13 input variables. The conclusion is therefore that the GA selected input variables as indicated in Table 1 provide the main contribution to the prediction of the target value. Here, not surprisingly, we observe contributions from pass line variations (via parameter *D_UP2_IMPOC_ENTRY*) and line speed (parameters *V_IMPOC_ENTRY* and *V_IMPOC_EXIT*), referring to the known influencing conditions on the IMPOC measurement [3, 5].

One of the input variables selected by the GA method is the *RED_PERC_TL* (= Reduction percentage at the tension leveller). This is remarkable, since the tension levelling process takes place only *after* the IMPOC_EXIT system. A relation with tension levelling seems not to be coincidental, since also the DT algorithm selects a tension levelling parameter, *I_TL* in this case, albeit, according to Fig. 4, its contribution is small. A short root cause study indicated that the tension levelling parameters are, via the recipe for this particular grade, linked to the hot rolling temperatures. As the hot rolling temperatures influence the microstructure and material properties, it forms a logical explanation why the tension levelling parameters can have a link with the IMPOC_EXIT measurement.

The second row for each model prediction in Table 2 indicates that when the IMPOC_ENTRY value is excluded from the input variables set, the error increases by 10–20%. Also Fig. 4 testifies the importance of IMPOC ENTRY for the final prediction. Nevertheless, it can be questioned if its contribution could be replaced by other, more upstream, process variables, like from the cold rolling or hot rolling processes, of course under the condition that these data are available to the HDG plant.

From the model predictions displayed in Fig. 6 and the errors listed in Table 2, we can deduce that more than 95% of the predictions are within a band of $\pm 2\ \sigma \approx \pm 200$ within the range of 3000 (i.e. 6500–9500) for the IMPOC_EXIT value. It was on purpose that we have selected a steel grade exhibiting a wide range in IMPOC values, to be able to develop and test the prediction model. Further studies will be needed with more data and with additional steel grades to validate our findings. In addition, for potential implementation of the developed knowledge, we will need to develop concepts for decision support systems and control systems which are inferred from the prediction models. Evidently, the robustness of these systems critically depends on the robustness with which the input data can be provided. In a harsh environment like a steel production plant, this is not always trivial, as testified by the heavy filtering of the original dataset required for the present analysis. Hence, the reliability in the data provision might be regarded as additional constraint to the selection of the set of input variables, especially when considering their use in real-time quality management and real-time process control. In the future the knowledge in terms of relation among process parameters and IMPOC signal highlighted in this work will be used for the development and validation of further AI-based models that use different sets of coils for training and validation purposes respectively. These tests will put into practice more advanced cross-validation techniques and data selection strategies in order to give a more precise insight of the exploitation of such approach during actual plant operation.

Acknowledgements. The work described in the present paper has been developed within the project entitled "Online Microstructure Analytics" (Ref. OMA, Grant Agreement No. 847296) that has received funding from the Research Fund for Coal and Steel of the European Union, which is gratefully acknowledged. The sole responsibility of the issues treated in the present paper lies with the authors; the Commission is not responsible for any use that may be made of the information contained therein.

References

1. Kroos, J., Stolzenberg, M., et al.: Combined Measuring System for an Improved Non-Destructive Determination of the Mechanical/Technological Material Properties, Technical Steel Research, final report, Office for Official Publications of the European Communities, Luxembourg (2005)
2. Midroit, F., Nogues, M., Damoiselet, F., Lindh-Ulmgren, E., Hutchinson, B., Ericsson, M., Morris, P.F.: New approaches to non-destructive characterisation of microstructure and applications to online control of steel quality (NANDACS), Contact Nr 7210-PR/94, EUR 21977, Final Report (2006), ISBN 92–79–01089–1
3. Stolzenberg, M., Schmidt, R., Casajus, A., Kebe, T., Falkenstrom, M., Martínez-de-Guerenu, A., Link, N., Ploegaert, H., van den Berg, F., Peyton, A.: Online material characterisation at strip production (OMC), Technical Steel Research, Publications of the European Communities, Luxembourg (2013) ISBN 978-92-79-28961-3 ISSN: 1831–9424, https://doi.org/10.2777/77313
4. van den Berg, F., Kok, P., Yang, H., et al.: In-line characterisation of microstructure and mechanical properties in the manufacturing of steel strip for the purpose of product uniformity control. In: Proceedings 19th World Conference on Non-Destructive Testing, WCNDT (2016)
5. van den Berg, F., et al.: PUC- Product Uniformity Control, RFSR-CT-2013-00031 (in press at the Publications Office of the European Union)
6. Matyuk, V.F., Delendikh, M.N., Osipov, A.A., Pinchukov, D.A.: The plant for pulse magnetic on-line testing of mechanical properties of rolled products. In: Proceedings Of the 9th European Conference on Non-Destructive Testing, ECNDT 2006 - Th.3.7.5. (2006)
7. Scheppe, D.: IMPOC - Increasing production yield by online measuremenent of material properties. SEAISI Q. (South East Asia Iron and Steel Institute) **83**, 41–46 (2009)
8. EMG Automation GmbH, Wenden, Germany. https://www.emg-automation.com
9. Skarlatos, A., Reboud, C., Svaton, T., de Guerenu, A.M., Kebe, T., Van den Berg, F.: Modelling the IMPOC response for different steel strips. In: Proceedings of the 19th World Conference on Non-Destructive Testing WCNDT 2016 (2016)

10. Skarlatos, A., Reboud, C., Kebe, T., Van den Berg, F., Martínez-de-Guerenu, A.: Intensive simulation tools dedicated to the monitoring of material properties. In: Proceedings of the 12th European Conference on Non-Destructive Testing ECNDT (2018)
11. Kebe, T., Dürr, W., Goureev, D., Reimann, G.: Einsatz des IMPOC-Verfahrens zur Bestimmung mechanisch-technologischer Eigenschaften von Stahlband. In: Proceedings of DGZfP Jahrestagung 2011, Mi.1.C.1. (2011)
12. Bärwald, M., et al: Online Quality Monitoring of a CGL, Steel Times International, Digital Edition Feb 2016, pp. 24-29 (2016)
13. van den Berg, F.D., et al.: Results of the European collaborative project "Product Uniformity Control" to improve the inline sensing of mechanical properties and microstructure of automotive steels. In: Proceedings of the 12th European Conference on Non-Destructive Testing ECNDT 2018 (2018)
14. Bärwald, M., Sommers, U., Biglari, M., Montagna, E., Beugeling, W., Lhoest, A.: Online quality monitoring in galvanizing lines. In: METEC & ESTAD conference (2015)
15. Nastasi, G., et al.: Chill marks effect detection algorithm for plant IMPOC data. In: Proceedings of the 19th World Conference on Non-Destructive Testing WCNDT 2016, (2016)
16. Mocci, C., Vannucci, M., Colla, V., Van Den Berg, F.D., Schmidt, R.: SOM analysis of IMPOC data in relation to process parameters in strip steel production. In: Proceedings of the 12th European Conference on Non-Destructive Testing ECNDT 2018 (2018)
17. Branca, T.A., Fornai, B., Colla, V., Murri, M.M., Streppa, E., Schröder, A.J.: The challenge of digitalization in the steel sector. Metals **10**(2), 288 (2020)
18. Brandenburger, J., Colla, V., Nastasi, G., Ferro, F., Schirm, C., Melcher, J.: Big data solution for quality monitoring and improvement on flat steel production. IFAC-PapersOnLine **49**(20), 55–60 (2016)
19. Sgarbi, M., Colla, V., Cateni, S., Higson, S.: Pre-processing of data coming from a laser-EMAT system for non-destructive testing of steel slabs. ISA Trans. **51**(1), 181–188 (2012)
20. Dimatteo, A., Vannucci, M., Colla, V.: Prediction of hot deformation resistance during processing of microalloyed steels in plate rolling process. Int. J. Adv. Manuf. Technol. **66**(9–12), 1511–1521 (2013)
21. Cateni, S., Colla, V., Vannucci, M.: A fuzzy system for combining different outliers detection methods. In: Proceedings of the IASTED International Conference on Artificial Intelligence and Applications, AIA 2009, pp. 87–93 (2009)
22. Cateni, S., Colla, V., Vannucci, M.: A fuzzy system for combining filter features selection methods. Int. J. Fuzzy Syst. **19**(4), 1168–1180 (2017)
23. Cateni, S., Colla, V., Vannucci, M.: General purpose input variables extraction: A genetic algorithm based procedure GIVE a GAP. In: ISDA 2009 - 9th International Conference on Intelligent Systems Design and Applications, pp. 1278–1283 (2009) Article No. 5364011
24. Cateni, S., Colla, V., Vannucci, M., Vannocci, M.: A procedure for building reduced reliable training datasets from real-world data. In: Proceedings of the IASTED International Conference on Artificial Intelligence and Applications, pp. 393–399. AIA (2014)
25. Galli, F., Ritacco, A., Lanciano, G., Vannocci, V., Colla, V., Vannucci, M.: Self-supervised pre-training of CNNs for flatness defect classification in the steelworks industry. Int. J. Adv. Intell. Inf. **6**(1), 13–22 (2020)
26. Colla, V., Vannucci, M., Bacchi, L., Valentini, R.: Neural networks-based prediction of hardenability of high performance carburizing steels for automotive applications. Metallurgia Italiana **112**(1), 47–53 (2020)
27. Hedar, A.-R., Ismail, R.: Hybrid genetic algorithm for minimum dominating set problem. In: Taniar, D., Gervasi, O., Murgante, B., Pardede, E., Apduhan, B.O. (eds.) Computational Science and Its Applications – ICCSA 2010, pp. 457–467. Springer, Berlin (2010)

Check for updates

Unsupervised Deep Learning for Detection of Non-uniform Surface Defect Distributions in Flat Steel Production

Nicolas Camargo-Torres(✉) and Jens Brandenburger

VDEH-Betriebsforschungsinstitut, Sohnstr. 65, 40237 Düsseldorf, Germany
{Nicolas.Camargo,Jens.Brandenburger}@BFI.de

Abstract. Over the past few years, deep learning architectures had achieved astonishing success in solving problems in a wide array of fields. The main requirement for the use of deep learning techniques is the existence of a large enough dataset for training. Recent advances in digitalization and Industry 4.0 have made it possible to acquire, process, and store large amounts of data, allowing the use of deep learning in industrial applications.

In the framework of a former research project a large database containing product and production data for the manufacture of flat steel was generated. This database contains, among other information, the in-coil positions of surface defects over multiple production stages. An analysis of these data suggested that the presence of an accumulation of surface defects at certain regions of the coil could be an indication of problems in the configuration of the production parameters. This prompted the need for a tool that can automatically identify production stages where a non-uniform distribution of surface defects is present in the coils.

This work uses a deep learning model to automatically detect and cluster coils with non-uniform defect distributions. Since for this purpose generally no class labels are provided in the databases, only unsupervised algorithms were used.

Keywords: Deep learning · Industry4.0 · Unsupervised learning · Clustering

1 Introduction

Nowadays, data-driven approaches are being intensively used in many areas of science and engineering, even in those where previously the classical analytical approach seemed to be highly satisfactory, such as mechanics and material science [1].

Deep learning models specifically had achieved astonishing success in solving problems in a wide array of fields including machine vision, pattern recognition, and material inspection [2–4].

Recently, deep learning has also been successfully used with data from surface inspection systems. In [5] and [6], convolutional neural networks (CNNs) are used for the detection of surface defects on labeled datasets, in [7] CNNs are used for the supervised classification of different types of steel defects, and in [8], an extension of a CNN accompanied by a max-voting scheme is used for steel microstructural classification.

V. Colla and C. Pietrosanti (Eds.): ESTEP 2020, AISC 1338, pp. 92–103, 2021.
https://doi.org/10.1007/978-3-030-69367-1_8

One of deep learning's main advantages is its capacity to extract meaningful features of the data on its own, the so-called feature engineering [9]. However, a major requirement for deep learning techniques to perform well is the existence of a large enough dataset for training [10].

The recent growing interest in Industry 4.0 and the application of its principles in manufacturing will allow the acquisition, storage, and processing of vast amounts of information from production plants, generating the so-called industrial big data [11].

Additionally, in recent years the computation power of computers has increased significantly. The component density on a computer chip continues to double every two years, as predicted by Moore's Law, and comparable doubling rates have been observed in chip speeds and computer speeds. It also seems that the conditions exist to keep sustaining such exponential growth in the years to come [12]. Significant advancements in modern graphic processor units (GPUs) and the use of distributed computing approaches have also contributed to increasing the performance of machine learning algorithms.

These factors have the potential to enable the extensive use of deep learning models in manufacturing, mainly with the goals of increasing quality, agility, and productivity. However, new reliable data-handling techniques are needed to effectively utilize such large amounts of data [11].

In this work, a deep learning clustering model was trained with the purpose of identifying distributions of surface defects with suspicious characteristics. The model was initially tested with an artificially generated dataset in order to obtain a better understanding of its working principle and to perform a controlled evaluation of the clustering results. Afterward, the model was tested with a new dataset generated with data from a real industrial application.

This paper is structured as follows. Section 2 details how the images of the distribution of surface defects are generated and explains the importance of identifying images with non-uniform distributions. Section 3 presents the characteristics of the deep learning clustering model used. Section 4 gives details about the generation of the artificial dataset and reports the clustering results obtained. Section 5 reports the results obtained on clustering a dataset with defect distributions from real industrial data. Section 6 concludes with a summary of the results obtained and provides an outlook for a possible future application.

2 Characterization of Surface Defect Distributions

The surface defect distributions referred to in this work originate from the multi-scale data representation model presented in [13]. This data model allows the storage of any information measured during the production process of a coil. Especially interesting are the stored data from the automatic surface inspection systems (ASIS). Using this data, the model can be used to correlate the position of the surface defects detected during production with other product and process information.

Another important characteristic of this model is that it allows the fast aggregation of data stored over multiple steel coils. This aggregation is possible because of the way the model stores the information. For each coil, the model associates the measured data into grids of constant length and width as shown in Fig. 1. For the generation of the distribution images, each rectangle $R_{x,y}$ of the grid in the coil is associated to a pixel $p_{x,y}$ of the image. Then, the brightness value of each pixel is set to the number of surface defects detected inside the corresponding rectangle $R_{x,y}$. This approach allows the generation of defect distribution images of a fixed size and allows a fast aggregation of spatial information.

Since the size of the grid can be configured, the data model also allows the configuration of the resolution of the image depending on the level of detail needed. The highest resolution currently stored by the system being images with 512 × 256 pixels.

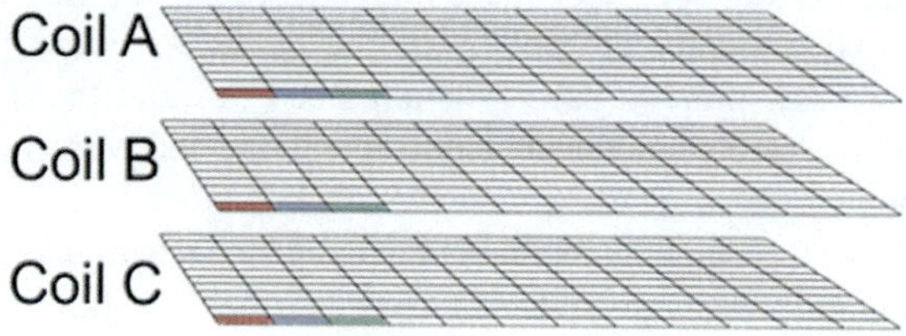

Fig. 1. Schematic view of the aggregation of multiple coils represented by the data model described in [13]

The motivation for exploring the characteristics of the defect distribution images of coils comes from one of the applications presented in [13] and [14]. There, the data model was used for tracking the position of one specific type of surface defect across multiples production stages during one year of production. The resulting distribution images after aggregating all the coils corresponding to two of the production stages are shown in Fig. 2. In these images, the darker blue a pixel is, the more surface defects were detected in the associated region of the coils. In [14] it was shown that the images presented in Fig. 2 can be used to interactively identify the root cause of the surface defects. In the case of the image on the right, where a clear accumulation of the defects is visible at the beginning of the coil, the position of the defects can be correlated with the causative factors of the defects. These results showed that the presence of non-uniformities in the distribution of defects could indicate problems in the configuration of the process parameters.

Thus, identifying when certain material types, production periods, or whole productions lines present such a noticeable defect distribution or a visible accumulation of defects over multiple coils, can help quality engineers to identify causes of defects and to solve quality issues.

However, today there is no tool available to identify these distributions automatically and the process of evaluating the coils is done manually. This process is very time-consuming and requires dedicated process know-how for the identification of the right composition of coils that leads to suspicious distributions. In [15] a decision tree was used to automatically find correlations with process parameters, but this was limited to a 1-dimensional analysis along the coil-length without considering the width distribution.

The automatic identification of coils with an accumulation of surface defects and a corresponding clustering of the coils with a quasi-uniform distribution would have a great potential for increasing quality supervision in steel production. The data model proposed in [13] allows a rapid retrieval of defect distribution images from different production stages. A tool able to automatically detect nonuniform defect distributions among these images could be used to support quality engineers in identifying groups of coils with quality deficiencies and identify root causes throughout production. In the present work, we propose the use of an unsupervised deep learning model to perform this clustering task.

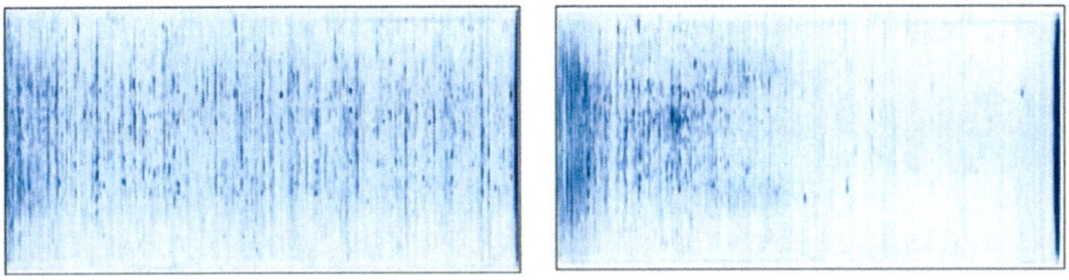

Fig. 2. Surface defect distribution images (left: uniform distribution, non-suspicious, right: accumulation of defects at the beginning of the coil, suspicious)

3 Deep Learning Clustering Model

Conventional clustering methods have issues on high-dimensional spaces and typically do not perform well on high-dimensional data [16]. Because of this, in the past few years, many researchers have used deep neural networks to facilitate the separation of data and to improve clustering performance.

For the selection of an algorithm for the clustering of defect distributions, we evaluated the state-of-the-art deep learning clustering models presented in [17]. We selected the Deep Embedding Clustering (DEC) algorithm described in [18]. This model was chosen due to its preferable properties related to the industrial requirements (performance efficiency, functional suitability, and robustness).

3.1 Deep Embedding Clustering (DEC)

DEC is a two-stage centroid-based method that uses a deep neural network both to learn a feature representation of the data and to generate cluster assignments.

The algorithm has two main objectives: (1) To learn a non-linear mapping function that transforms the input data from a high-dimensional data space X into a low-dimensional feature space Z. (2) To improve iteratively and simultaneously both the feature representations and the clustering assignments using an auxiliary target distribution derived from soft cluster assignments. The DEC algorithm consists of two phases: (1) parameter initialization using a deep autoencoder and (2) parameter optimization. For the optimization, the clustering assignments are refined by minimizing their Kullback-Leiber (KL) divergence to an auxiliary target distribution. The details of these two phases are presented below.

Phase 1: Autoencoder Training for Parameter Initialization. In the first stage, the DEC algorithm seeks to learn a non-linear mapping function $f_\theta : X \to Z$. For this purpose, DEC uses a stacked autoencoder (SAE), which is an autoencoder that uses a stack of multiple sparse autoencoders as hidden layers.

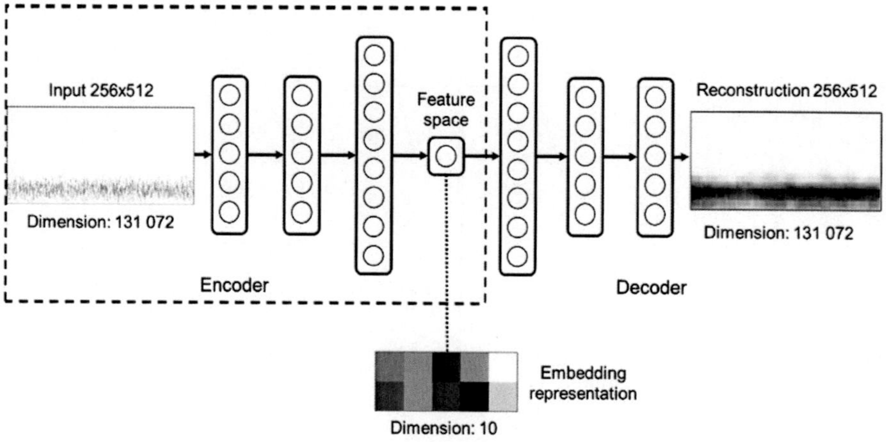

Fig. 3. Nonlinear mapping of the embedding representation using an autoencoder

DEC uses a fully connected architecture, in which each layer is a denoising autoencoder that tries to reconstruct the output from the previous layer after a random corruption. The final architecture of the network consists of a multilayer deep autoencoder with a bottleneck coding layer in the middle. The encoder part progressively extracts the features from the data and the decoder part is configured to mirror the structure of the encoder. The dimensionality of the feature space is given by the number of units used in the bottleneck layer of the autoencoder. The information encoded in this layer is referred to as the embedding representation of the data. Figure 3 presents the structure of the autoencoder network used for the extraction of the embeddings.

After the training of the autoencoder, the decoder part of the network is discarded and an additional clustering layer is stacked on top of the pre-trained encoder. Finally, to complete the initialization process, the k-means algorithm is applied to the embedding representation of the data to obtain the initial cluster centroids.

Phase 2: Parameter Optimization with KL Divergence. The second stage of the DEC algorithm consists of an iterative process in which the cluster assignments are gradually improved. This process alternates between two steps: (1) The algorithm computes a soft cluster assignment for the embedding representation of the data. (2) The algorithm updates all the parameters of the network and refines the cluster centroids. For this purpose, the network uses an auxiliary target distribution that allows it to learn from its high confidence assignments.

The soft cluster assignment q_{ij} corresponds to the probability of assigning a data sample i to a cluster j. DEC computes this probability using the Student's t-distribution to measure the similarity between the embeddings and the cluster centroids.

Figure 4 shows the final structure of the DEC network and presents an example of the soft cluster assignments. DEC defines an auxiliary target distribution p_{ij} to iteratively improve the cluster assignments. The objective of the network is to force its soft cluster assignments q_{ij} to be similar to the target distribution p_{ij}. The target distribution is designed to give the data points with a high confidence assignment a greater influence during the training process. The distribution is also normalized to prevent a distortion of the feature space by large clusters. DEC computes p_{ij} by squaring the soft assignment q_{ij} and then normalizing it by frequency per cluster.

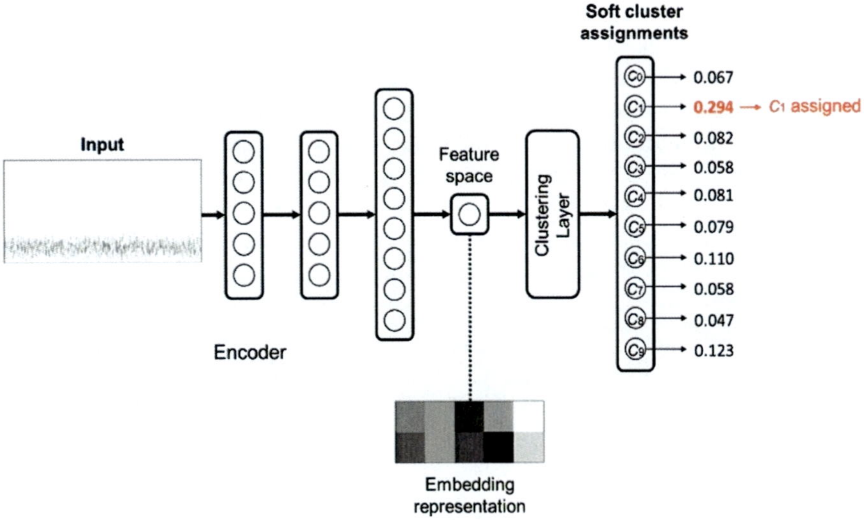

Fig. 4. DEC network structure.

During training, the DEC algorithm uses the KL divergence loss to match the soft assignment to the target distribution. The KL divergence is a measure of how much a probability distribution differs from another and is defined by:

$$L = KL(P||Q) = \sum_i \sum_j p_{ij} \log \frac{p_{ij}}{q_{ij}} \tag{1}$$

The gradients of the loss function L with respect to each embedded point are passed down the network using backpropagation to generate the parameter gradients. The DEC network then jointly optimizes the non-linear mapping function and the cluster centroids using Stochastic Gradient Descent (SGD).

4 Clustering of an Artificial Defect Distribution Dataset

In order to establish the viability of the use of the DEC model for the clustering of defect distributions, an evaluation of the accuracy of the model with a defect distribution dataset is needed. This evaluation requires previous knowledge of the true labels of the distributions present in the dataset. In the case of real industrial data, these labels are typically not available.

For this reason, in our approach, we propose the generation of an artificial dataset of defect distributions for the evaluation of the deep learning clustering model. Since we generate the data, the true labels of the distribution images are known and can be used for the evaluation of the results. Furthermore, the only limitation for the amount of data that can be generated is the disk space available. Thus, enough training data can be generated to allow the training of the deep learning model.

4.1 Generation of the Artificial Distribution Dataset

For the generation of the artificial distribution images suitable probability distributions were used, i.e. the uniform distribution and the normal distribution. We used a combination of these two distributions to simulate realistic distributions of surface defects found in coil manufacturing. We defined four classes of defect distributions for the artificial dataset: class 0 was defined as coils with a random distribution of defects, class 1 was defined as coils with a point-shaped accumulation of defects, class 2 was defined as coils with a horizontal stripe distribution of defects and class 3 was defined as coils with a vertical stripe distribution of defects.

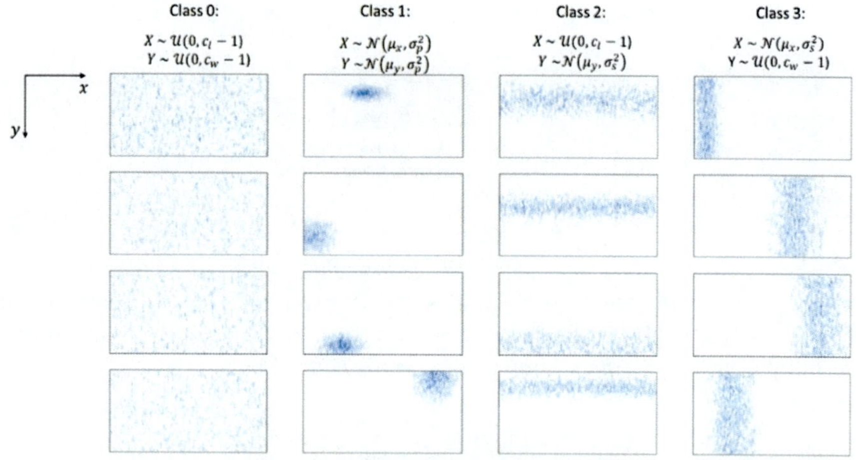

Fig. 5. Samples generated for the four classes of the artificial defect distribution dataset

Since in the case of real defect distribution images the position of the accumulation of defects on the coil can highly vary. For classes 1–3 we randomly changed the position of the accumulation of defects for each image. Figure 5 shows some examples of the generated samples for each artificial defect distribution class. The characteristics of the artificial images are similar to the images generated by the data model proposed in [13] in the resolution stage 8. The image size of the artificial images in pixels was therefore set to 512 × 256.

We generated two sets of data: a training dataset and a test dataset. For the training dataset, we generated 5000 samples for each distribution class for a total of 20000 training samples. For the test dataset, we generated 1000 samples for each distribution class for a total of 4000 test samples. We normalized all sample images so that, for each image, the maximum pixel value was 1 and the minimum pixel value was 0.

4.2 Network Characteristics

Although in the original implementation of the DEC model a fully connected architecture was used, we used a convolutional architecture for the autoencoder instead. The working principle of this autoencoder remains the same. This network is used for the dimensionality reduction of the input data. The advantage of this architecture, however, is that it is computationally more efficient since a fewer number of parameters need to be learned. We refer to this model as DEC Conv. Its architecture is presented in Fig. 6.

For the convolutional autoencoder, we used 16, 32, and 64 filters for the layers of the encoding part. The decoder was configured symmetrical to the encoder. For all the convolutional layers we used a kernel size of (3, 3). For all the max-pooling layers we used a window size of (2, 2). The embedding dimension was set to 10. We used the same number of neurons for the clustering layer of the model as the number of ground-truth labels, in this case, 4.

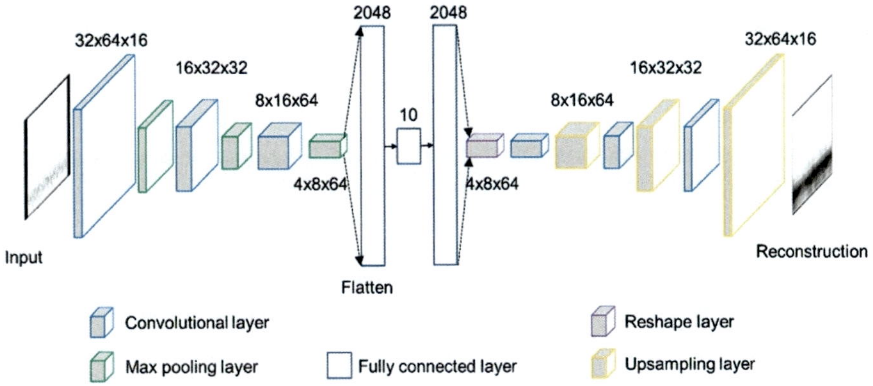

Fig. 6. Convolutional architecture of DEC

For the optimization, we used Stochastic gradient descent (SGD) and the learning rate was obtained using the learning rate finder (LRF) algorithm proposed in [19]. We also used multiple pre-processing techniques to guarantee the convergence of the training of the model.

4.3 Clustering Results

The DEC Conv model achieved a clustering accuracy of 99.9% in both the training and the test dataset. The resulting confusion matrix along with the PCA visualization of the generated embeddings are presented in Fig. 7. The PCA of the embeddings clearly shows that the four classes are mapped in different regions of the cluster space.

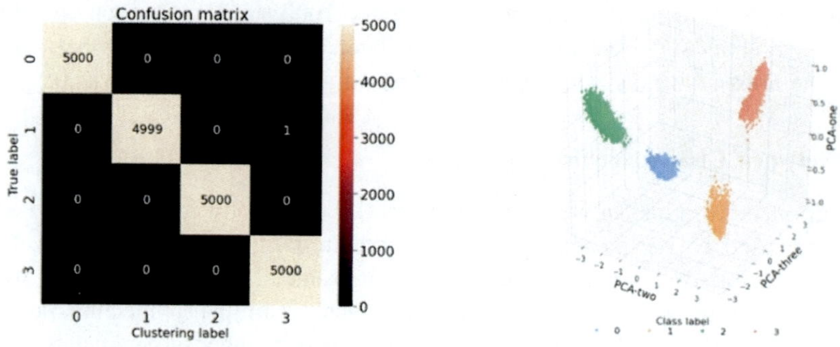

Fig. 7. Clustering result for artificial defect distribution images (left: Confusion matrix, right: Visualization of cluster space after DEC Conv model training)

5 Clustering of a Real Defect Distribution Dataset

Since the deep learning clustering model obtained very good results with artificial data, we tested it next with a defect distribution dataset generated from real industrial data.

For this dataset, we used the coils information obtained in [13]. More specifically the coils corresponding to the production stages with the distribution images presented in Fig. 2. These two distribution images have notably different characteristics. The image on the left has an almost uniform distribution of defects along the length of the coils and was generated by aggregating a total of 2171 coils, while the image on the right presents a clear accumulation of defects at the beginning of the coil and was generated by aggregating the defects from 1646 coils.

Since for each coil the corresponding production line was known, it is possible to use this information for the generation of labels for the evaluation of the clustering algorithm. The coils associated with the production stage with a uniform distribution were labeled as class 0 and the coils associated with the production stage with a non-uniform distribution were labeled as class 1.

5.1 Generation of the Real Distribution Dataset

Although the defect distributions can be clearly differentiated once the total amount of coils of each production stage are aggregated, if the coils are considered individually, in most cases, there is not a sufficient number of defects present for a clear distribution to be visible. In consequence, an important parameter that needs to be selected is the number of coils that must be aggregated for the generation of each distribution image.

We decided to use a set of 100 coils for the generation of each image since at this point the expected distributions for the two classes began to be clear. Figure 8 presents some examples of the resulting distribution images for both classes. To generate a number of samples large enough for the training of the deep clustering model, we selected a random combination of 100 coils for the generation of each distribution image. Using this approach, each coil can be used to generate multiple distribution images. Moreover, due to the random aspect of the selection, different distribution images with similar characteristics are generated.

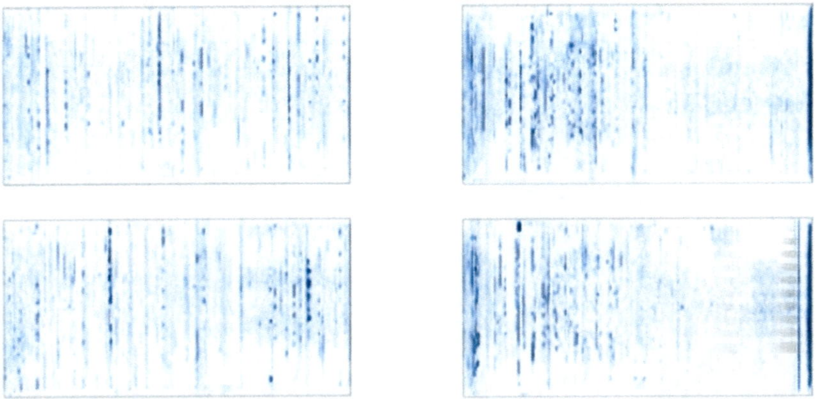

Fig. 8. Defect distribution images generated by aggregating the defects of sets of 100 coils (left: Coils form production stage with uniform distribution, right: Coils from production stage with non-uniform distribution)

Using this method, we generated 4000 distribution images for each class for the training dataset, for a total of 8000 training samples. For the test dataset, 1000 distribution images were generated for each class for a total of 2000 test samples.

5.2 Clustering Results

The network used had the same characteristics as the network presented in Sect. 4.2. Initially, we used a random initialization for the network and trained the model using only the data generated in Sect. 5.1.

The model was trained 10 times using different random seeds for the initialization. It was noted that the clustering results widely varied among the different initializations. Although some models were able to correctly cluster most of the data, in two out of the ten cases the model resulted in a clustering accuracy lower than 60%, the minimum being 57.31%. It was found that the problem stems from the autoencoder training used for the parameter initialization of the DEC Conv model. In the cases with low clustering accuracy, the training of the autoencoder did not converge and the network could not reconstruct the input images.

To reduce the initial random effect of the autoencoder training and improve the robustness of the results, the use of a pre-trained network for the initialization of the autoencoder is proposed. We used the weights of the autoencoder previously trained with artificial images from Sect. 4 for the initialization of the autoencoder used with real data. Afterward, the autoencoder was trained with the images of the real distribution dataset.

Using this approach, we trained the model 10 times again. In all cases, the model achieved a clustering accuracy higher than 98% both for the training and the test datasets. On average a clustering accuracy of 98.97% was achieved. Figure 9 presents the confusion matrix obtained and the PCA visualization of the embeddings for the training dataset.

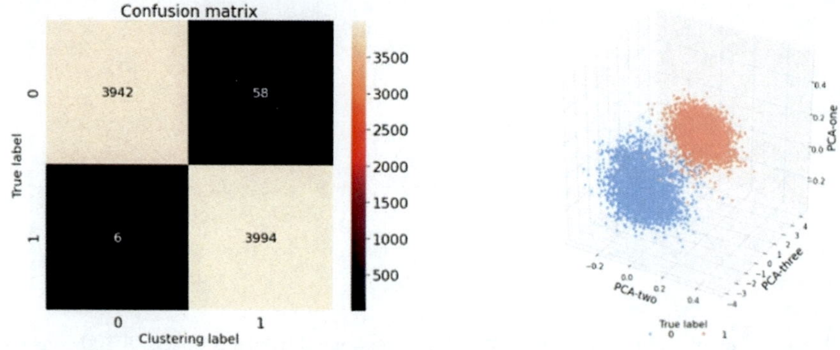

Fig. 9. Clustering results with real defect distribution images (left: Confusion matrix, right: Visualization of cluster space after DEC Conv mode training)

6 Conclusions

This paper presents the use of an autoencoder-based deep learning model for the clustering of surface defect distribution images from steel coils. The model was tested both with an artificially generated dataset and with a real defect distribution dataset.

In the case of the artificial dataset, the algorithm was able to correctly cluster more than 99.9% of the samples. The results show that the model was able to identify and differentiate the artificial distributions regardless of the position of the accumulation of the defects.

In the case of real surface defect distributions, the algorithm achieved on average a clustering accuracy of 98.97%. Moreover, the results show that using the weights of a network pre-trained with artificial data for the initialization of the autoencoder results in an improvement of the robustness of the clustering.

This clustering model can be used in future work for the online detection of suspicious defect distributions at production lines and help improve production quality.

References

1. Montáns, F., Chinesta, F., Gómez-Bombarelli, R., Kutz, J.: Data-driven modeling and learning in science and engineering. Compt. R. Mécanique **347**(11), 845–855 (2019)
2. Krizhevsky, A., Sutskever, I., Hinton, G.: ImageNet classification with deep convolutional neural networks. Adv. Neural. Inf. Process. Syst. **25**, 1097–1105 (2012)
3. Yu, J., Zheng, X., Wang, S.: A deep autoencoder feature learning method for process pattern recognition. J. Process Control **79**, 1–5 (2019)
4. Li, L., Ota, K., Dong, M.: Deep learning for smart industry: efficient manufacture inspection system with fog computing. IEEE Trans. Industr. Inf. **14**(10), 4665–4673 (2018)
5. Arikan, S., Varanasi, K., Stricker, D.: Surface Defect Classification in Real-Time Using Convolutional Neural Networks. arXiv e-prints (2019)
6. Lee, S., Tama, B., Moon, S., Lee, S.: Steel surface defect diagnostics using deep convolutional neural network and class activation map. Appl. Sci. **9**(24), 5449 (2019)
7. Masci, J., Meier, U., Ciresan, D., Schmidhuber, J., Fricout, G.: Steel defect classification with Max-pooling convolutional neural networks. In: The 2012 International Joint Conference on Neural Networks (IJCNN), pp. 1–6. IEEE, Piscataway (2012)
8. Azimi, S., Britz, D., Engstler, M., Fritz, M., Mücklich, F.: Advanced steel microstructural classification by deep learning methods. Sci Rep **8**, 2128 (2018)
9. Patterson, J., Gibson, A.: Deep Learning: A Practitioner's Approach, 1st edn. O'Reilly, Beijing (2017)
10. Bach, S., He, B., Ratner, A., Ré, C.: Learning the structure of generative models without labeled data. In: Proceedings of the 34th International Conference on Machine Learning, pp. 273–282. Sydney, Australia (2017) JMLR.org
11. Sharma, A., Pandey, H.: Big data and analytics in industry 4.0. In: Nayyar, A., Kumar, A. (eds.) A roadmap to Industry 4.0. Advances in Science, Technology & Innovation, IEREK Interdisciplinary Series for Sustainable Development, pp. 57–72. Springer, Cham (2020)
12. Denning, P., Lewis, T.: Exponential laws of computing growth. Commun. ACM **60**(1), 54–65 (2016)
13. Brandenburger, J., Colla, V., Cateni, S., Vignali, A., Ferro, F., Schirm, C., Melcher, J.: Applying big data concepts to improve flat steel production processes. In: Roy, S., Samui, P., Deo, R., Ntalampiras, S. (eds.) Big Data in Engineering Applications, pp. 1–20. Springer, Singapore (2018)
14. Brandenburger, J., Schirm, C., Melcher, J.: Instant interactive analysis - how visualisation can help to improve product quality. SIS.EUROPE - Surface Inspection Summit, Aachen, Germany (2016)
15. Brandenburger, J., Colla, V., Nastasi, G., Ferro, F., Schirm, C., Melcher, J.: Big data solution for quality monitoring and improvement on flat steel production. In: 17th IFAC Symposium on Control, Optimization and Automation in Mining, Mineral and Metal Processing (MMM), pp. 55–60 (2016)
16. Hinton, G., Sejnowski, T.: Unsupervised Learning: Foundations of Neural Computation. Computational Neuroscience. MIT Press, Cambridge (1999)
17. Min, E., Guo, X., Liu, Q., Zhang, G., Cui, J., Long, J.: A survey of clusteringWith deep learning: from the perspective of network architecture. IEEE Access **6**, 39501–39514 (2018)
18. Xie, J., Girshick, R., Farhadi, A.: Unsupervised deep embedding for clustering analysis. In: 33rd International Conference on Machine Learning, pp. 478–487. PMLR, New York (2016)
19. Howard, J., Gugger, S.: Fastai: a Layered API for deep learning. Information **11**(2), 108 (2020)

Machine Learning-Based Models for Supporting Optimal Exploitation of Process Off-Gases in Integrated Steelworks

Ismael Matino[1(✉)], Stefano Dettori[1], Angelo Castellano[1], Ruben Matino[1], Claudio Mocci[1], Marco Vannocci[1], Alessandro Maddaloni[1], Valentina Colla[1], and Andreas Wolff[2]

[1] Scuola Superiore Sant'Anna, TeCIP Institute, ICT-COISP Center, Pisa, Italy
{i.matino, s.dettori, a.castellano, r.matino, c.mocci, m.vannocci, a.maddaloni, v.colla}@santannapisa.it
[2] Betriebsforschungsinstitut, Sohnstrasse 65, 40237 Düsseldorf, Germany
andreas.wolff@bfi.de

Abstract. Within integrated steelworks, several sub-processes produce off-gases, which are suitable for reuse as energy sources for other internal processes as well as for the production of energy. An adequate and optimal distribution of these gases among their users allows valorizing at best their energy content by minimizing the need to both burn them through torches due to storage issues and to acquire natural gas to satisfy the internal energetic demand. To this purpose, the volume and energetic value of produced gases as well as the demands from internal users must be known in advance, in order to implement model-predictive control strategies aimed at satisfying the demands on the short-medium term based on the production scheduling. Such forecasting knowledge also enhances the capability to react to the variability of the process scheduling as well as to other unforeseen events.

The paper depicts an application of Machine Learning-based models to forecast off-gases and further energy carriers productions and demands within integrated steelworks. The forecasting models are integrated into a complex hierarchical control strategy aimed at optimizing the distribution of such gases.

Keywords: Decision support system · Echo state neural network · Model predictive control · Process off-gas management · Steelworks sustainability

1 Introduction

Steel has always been and will continue to be a fundamental material for the society thanks to its characteristics of durability, versatility and recyclability. It plays a central role for the society development as well as for its evolution toward an environmental friendly community. To keep such role, the European steel sector attributes a fundamental importance to resource and energy efficiency and to reduction of carbon footprint and emissions related to its production processes. The European steel production represents about 16% of worldwide production and 8.5% belongs to the European Union (EU) countries [1]. Considering these figures, improving the environmental and

V. Colla and C. Pietrosanti (Eds.): ESTEP 2020, AISC 1338, pp. 104–118, 2021.
https://doi.org/10.1007/978-3-030-69367-1_9

economic sustainability of the steel sector can help increase the competitiveness of European industry and improving the European economy as a whole. This is considered fundamental by the EU that through the *European Green Deal* will "*support industry to innovate and to become global leaders in the green economy*" [2].

In this context, integrated steelworks are focusing their efforts toward reduction of emissions as well as energy and resource consumptions. The integrated steelmaking route actually exploits a high amount of resources and generates important volumes of by-products, wastes and emissions. In addition, about 25% of production costs are related to energy [3]. Often resource and energy management are linked together and an improvement of resource management can lead to an optimization of energy management and vice versa. Process Off-Gases (POGs) are exemplar in this sense, as they are a by-product but also a resource, being a significant energy carrier.

POGs are produced in main production steps of the integrated route: Coke Oven Gas (COG) derives from coke production, Blast Furnace Gas (BFG) is produced together with pig iron and Basic Oxygen Furnace Gas (BOFG) is generated during pig iron conversion into steel. These gases are rich in CO and H_2, have a significant Net Calorific Value (NCV) and, therefore, are generally recovered and exploited as internal energy source or for electricity production. Nevertheless, their management and distribution are often not optimized due to the complexity of the integrated route that combines different steps and several units, of the related gas/steam/electricity networks and also of the involved production plans. Gas over-production and under-production are common issues which lead, respectively, to storage issues (i.e. full gasholder is often translated in flared gas or slowdown of the production) or to dissatisfaction of gas demand with consequent exploitation of natural gas.

Different solutions can be implemented to improve the use of these valuable resources avoiding their waste or their lack. Some solutions are related to Carbon Capture and Usage and Storage (CCUS) or other chemical conversion-related solutions. For instance, the *Steelanol* project aims at converting the CO and H_2 contained in the BFG into ethanol by using fermentation by microbes [4]. The *Carbon2Chem* project promotes the use of steelmaking top gases for the production of basic chemicals, plastics and synthetic fuels [5, 6]. The ongoing *i3upgrade* project focuses on the production of methane and methanol by exploiting steelworks gases after enrichment with green hydrogen and by coupling innovative processes and an advanced dispatch controller [7]. The use of steelworks gases for methanol production is also investigated in other literature studies [8–11]: different synthesis processes are investigated and techno-economic analyses are carried out. The *VALORCO* project investigates the capture of carbon dioxide of BFG through amine scrubbing technologies [12].

On the other hand, different investigations and studies focuses on improving the management of process off-gases and maximizing their internal use. To this aim, forecasting models have also been developed to support optimization approaches. For instance, Neural Networks (NN)-based models forecasting the production of BFG were developed by Zhang et al. [13] and by Yang et al. [14], who exploits genetic algorithm and least-squares support-vector machine. Zhao et al. [15, 16] uses the novel Echo State NNs (ESN) for both BFG and COG production. ESNs are also used in [17] to forecast steam production and consumption within steelworks. Porzio et al. developed an offline Decision Support System (DSS) including modules for static simulation and optimization allowing the monitoring and the reduction of CO_2 emissions from wasted off-gases by providing offline optimal solutions of gases distribution [18–20]. Different multi objective optimization approaches were investigated and compared for optimal distribution of off-gases [20]; among others, interesting are Genetic Algorithm [21], Mixed Integer Linear Programming (MILP) [22, 23] and multi-period optimization [24]. However, all of these works present some issues that can be listed as follows: the gas networks (one for each gas), the steam network and all involved processes, phenomena and related operating conditions and constraints are not considered as a whole and the different synergetic aspects are neglected; not all the energy carriers are considered jointly; the volume and energy amount of the different energy carries are sometimes decoupled; production plan is not considered in the forecasting and optimization purposes; if global optimization is carried out, it is made in offline mode. Thus, starting from the high valuable works found in literature and by considering their bottlenecks, a DSS has been developed during a cutting-edge project, namely *GASNET*. The DSS aims to fill the previous listed gaps and to obtain a holistic and online optimization of the integrated steelworks off-gas management by coupling in its core a forecasting model library and an advanced online optimization tool. In particular, the model library includes, among others, Machine Learning (ML)-based models that allow the prediction of production and demand of amount and energy content of main energy carriers (i.e. off-gases, steam and electricity). On the other hand, a Hierarchical Control and Supervision system based on Economic Model Prediction Control (EMPC) methodologies is applied for the optimization of POGs distribution and energy flows.

The work presented in the paper improves the current state of the art concerning off-gas management in the steel field by introducing the consideration of the whole production plant, through an approach involving a sort of "digital twin" of the entire steelworks from the point of view of energy production, exploitation and transformation. This digital twin, based on different ML methodologies, has been integrated into an advanced control system, which also improves the current state of the art for the optimization of energy flows within integrated steelworks.

The paper is organized as follows: Sect. 2 summarizes the aims of the GASNET project and introduces the developed software. Section 3 describes the adopted modelling approaches and optimization strategies. Section 4 overviews the most representative models. Section 5 discusses the simulation and optimization results. Finally, the Sect. 6 provides some concluding remarks.

2 Overview of the GASNET Project and Software

GASNET is a project funded by the Research Fund for Coal and Steel, which aimed at upgrading the actual management of off-gases coming from the steelmaking processes by optimizing the distribution and use of these gases. This was obtained thanks to the holistic view of all involved "actors" and related constraints in the energy management. The main outcome is an integrated simulation and decision support tool, which is articulated in 2 modules that can be installed and run independently or can be jointly used:

1. An online module that allows continuous monitoring and optimization of energy carriers networks, which is in turn composed of the following sub-modules: industrial plant and energy database, forecasting model library, online control and optimization tool, Key Performance Indicator (KPIs) calculator.
2. An offline network optimizer which allows also making structural modifications (e.g. dismantling or building new pipes) [25].

Both modules have an easy-to-use Graphical User Interface (GUI) that allows the exploitation of the different software areas (e.g. model configuration, model use, optimization configuration) and the visualization of the main results. Some screenshots of the GUI of the two modules are depicted in Fig. 1.

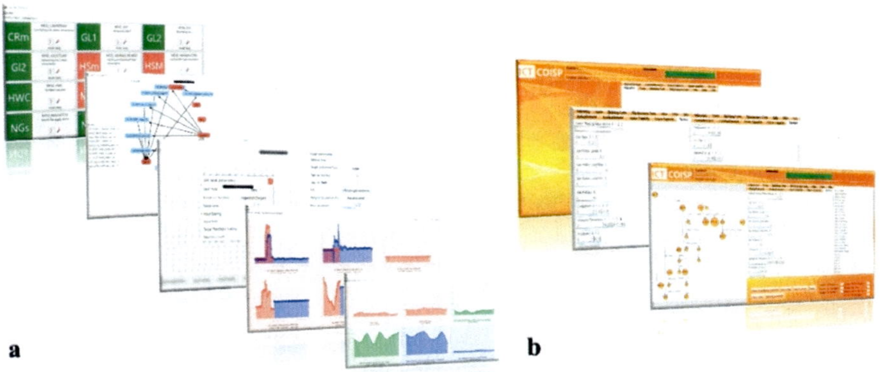

a b

Fig. 1. GUI of the GASNET software: a. online module; b. offline network optimizer.

3 Materials and Methods

Forecasting energy carriers production and demand is fundamental to implement control and optimization strategies to satisfy the demands on the short-medium terms by considering the production scheduling. A correct selection of modelling and optimization approaches is fundamental to rich the goal. Acquisition of process know-how through a strict cooperation with plant staff and literature analysis were the starting point. The selected methodologies are discussed in the following sub-paragraphs.

3.1 Modelling Methodologies

The aim of forecasting models is to virtually reproduce the dynamic of the different processes, phenomena and unit operations involved in the production and use of the energy carriers in the integrated steelworks and to continuously forecast their behavior in a time horizon of 2 h, which is considered sufficient for short-term optimization purposes.

A first distinction needs to be done between producers and users. Some processes depend on their own evolution through time, while other ones are strictly correlated to production plan or process data or their behavior is difficult to predict due a strong link with human decision-making. Most of the processes present complex dynamics and non-linear behavior. Therefore, a unique general-purpose modelling methodology is not suitable to fit all the considered process/phenomena. Moreover, a correct selection of the inputs is fundamental, taking into account staff experience, process knowledge, availability of continuous monitored and registered data. Therefore, inputs and modelling approaches have been accurately and carefully selected on a case-by-case basis exploiting literature and preliminary tests. Regarding the inputs selection, after a collection of high amount of available data from historical databases of monitored variables, detection and removal of outliers was first carried out [26–28]. Then feature space reduction was done in two stages for the datasets related to the different process/phenomena to be modelled: detection and elimination of redundant variables and selection of variables, which mostly affect the considered target [26, 29, 30].

On the other hand, concerning the modelling techniques, literature and authors previous experience suggested advantages and drawbacks of different methods physical and chemical based or data-driven. However, a preliminary testing phase was carried out to verify the applicability of different modelling approaches, their accuracy, computational burden and response for each process/phenomena to be represented. In other words, firstly the possibility to apply physical/chemical based approaches was assessed. However, in many cases, the lack of relevant information prevented both accurate development of dynamic models and systematic validation considering different status of the modelled process. In addition, the computational burden and time required by this kind of models was not suitable for continuous/real-time simulation and optimization purposes due to the high number of inputs/outputs and, in some cases, of complex iterations. Afterwards, the exploitation of a mixed approach was attempted according to the authors' previous works [18]. Such approach combines flowsheeting and ML/interpolation approaches. Stationary flowsheeting models were developed, tuned and validated by mixing real and literature data and by using well-known commercial simulation software and allow reproducing the physical/chemical phenomena of a process in different scenarios. These models are generally solved following a sequential modular approach and the flowsheet needs to converge at every iteration, because it computes every unit operation in sequence by leading to unsuitable times for the optimization task. However, they can be used as generators of data to be fed to dedicated NNs for approximate the flowsheeting simulations [31] and obtaining NN-based models with negligible computational time. This approach showed good results but it has two main issues: the propagation of errors and the difficulty to reproduce the dynamic of the systems (this feature can be neglected for off-line optimization but not for the real-time one) especially in case of complex dynamic processes, as a stationary model is used to train the NNs. Thus, it was decided to check different "native" black box techniques, especially for complex processes in order to exploit their potential in terms of capturing dynamic features, using lower numbers of inputs and requiring lower computational burden and time. After some tests, only for simpler processes equation-based models were used, but most of the models are based on ML approaches. In particular, for less state-dependent dynamics, Feed Forward NNs (FFNN) were used, with a multi-layer perceptron topology; while, for complex non-linear time dependencies, the potential of ESN was exploited. ESN represents a novel, powerful and effective method to model complex nonlinear time dependencies [15, 16, 32–36]. In this work, they were used both in its shallow and deep architecture: they differ from the number of "dynamic" layers, namely "reservoirs"; in particular shallow ESN is provided of a single reservoir, while Deep ESN (DESN) is provided of different reservoirs in series [37, 38]. The reservoir is the nucleus of ESN, as it generates a rich set of dynamics that are combined to make regression on a target by means of supervised algorithms and thus to reconstruct complex dynamical relationship between inputs and outputs. The particular features of the dynamic reservoir provide it with an auto-feedback nature that is translated in a short-term memory of past dynamic states. Another advantage of ESNs lies in the fact that, although they are recurrent NNs, they do not exploit iterative routines and thus they have a low computational training cost.

For each ML-based model, the global dataset of real industrial data, corresponding to several months of production, were divided in two sets: 60÷70% of the data were used for training, the other ones for validation. The architecture setup of the models was optimized through a proper selection of the hyper-parameters of the specific adopted topology. In particular, for the ESN-based approach the total number of neuron, the number of layers, spectral radius, input scaling and Tikhonov regularization parameter were considered. The FFNNs architecture was optimized in terms of number of layers and total number of neurons. The mentioned hyper-parameters were optimized through a random search algorithm aimed at increasing the model accuracy, tested through the Normalized Root Mean Square Error (NRMSE):

$$NRMSE = \frac{\sqrt{\frac{1}{S}\sum_{i=1}^{S}(Fo_i - o_i)^2}}{M - m} \cdot 100 \tag{1}$$

where S is the data number, Fo_i and o_i are, respectively, the i-th sample of the forecasted output and real target, M and m are maximum and minimum real target values.

3.2 Optimization Strategies

After data preparation through synchronization of different datasets, deletion of duplicates, outlier detection and elimination through the techniques, which are described in the previous Section, the analysis of the collected data related to the plant management, POGs and steam networks structure can reveal several issues. The different behaviors of energy production, transformation and transfer (in terms of different involved energy carriers) and the production plan are considered and compared. Firstly, the overall plant production scheduling is usually not optimized. This involves energy flows that highlight large differences between the peaks of energy production and transformation and its consumption. This imbalance must be managed, firstly, through gasholders (GH) acting as buffers and allowing a margin of safety and controllability of the system, secondly by an intelligent scheduling of electricity production in internal power plants, the main POG consumers. The second evident bottleneck is the size of the GHs, whose dimensions guarantee a buffer of 30–90 min. Optimizing the use of POGs means acting in the short term, with a 2-h prediction horizon to ensure a sufficient controllability margin.

An optimal scheduling of energy flows is important to provide general indications to plant managers and process operators for improving the production scheduling. In this case, the supervision and control strategy must give indications in the medium and long term. Therefore, the supervision and control system must be structured in a hierarchical manner, according to the time horizon in which the plant energy flows can be optimized. The core of the developed DSS is based on a two-level hierarchical optimization system shown in Fig. 2.

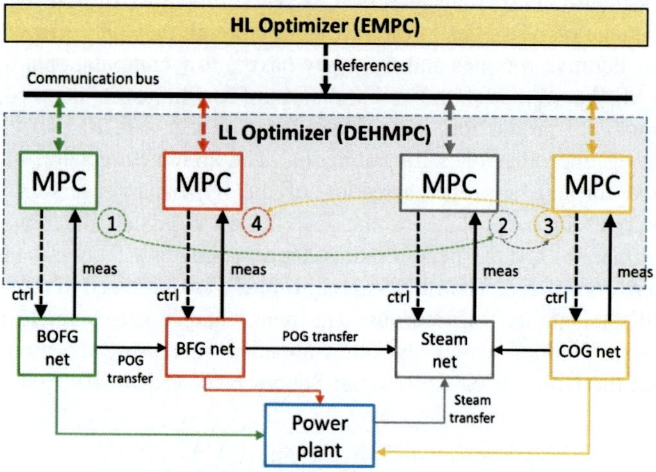

Fig. 2. Structure of the Supervision and control system.

The first level (*HL Optimizer*) optimizes the energy flow in the medium and long term, with a prediction interval between 1 day and 1 week, which allows formulating a set of possible control actions that mainly aim at scheduling the POG use in the power plant. The control action output by the *HL Optimizer* is used by the *LL Optimizer* that allows distributing the energy flows in the short term, with a higher control frequency.

The *HL Optimizer* implements a Linear Programming formulation within a control strategy based on the Economic Model Predictive Control (EMPC). The control action is computed every 15 min by minimizing the management cost in terms of consumption of Natural Gas (NG) and purchased electricity and the environmental costs related to the disposal of excess POG through torches. The optimization takes into account the main limits of the POG, steam and electricity networks, in terms of minimum and maximum operating points (e.g. gasholder levels, produced energy in the power plant, etc.).

The *LL optimizer* aims at distributing the energy flows over a prediction and control horizon of 2 h, with a control frequency of 1 min. This allows managing the dynamics of the systems involved and, above all, considering the real operating limits of each equipment. More in detail, the *LL optimizer* is structured as a distributed controller that implements several Local Economic Hybrid Model Predictive Controllers (LEHMPC), by solving in real-time a MILP formulation for each specific network [39]. Each energy carrier network (BOFG, BFG, COG and steam) is sequentially optimized. In particular, Fig. 2 shows a possible numerated sequence of the scheduling of each network MPC and their mutual and sequential communication through the bus. The communication with the DSS (and its modules) is carried out through an interface layer between the DSS and the IT system of the plant established with ProDISS, a databank-based communication interface, which exchanges messages with different systems (e.g. plants). The communication procedure is based on different steps: the ProDISS-DSS interface receives a message from a plant with required data and stores them in an input layer in the database (DB). The communication interface starts with a PL/SQL package, which, after reading the data from the input layer, provides them in the structure required by the DSS. The data are stored in another separate layer and the DSS reads them. In this way, the whole application can run on several machines.

The supervision and control system exploits the presented model library as disturbance predictor and includes an additional set of linear models describing the dynamics of the controlled system with high accuracy, such as steam boilers, steam accumulators, GHs, power plant and turbines. The system has been prototyped in Matlab environment through YALMIP Toolbox and translated in C++ through the optimization libraries of the Google OR Tools. The DSS does not implement a closed loop control strategy, but provide advices to all the operators of the steam, POGs and power plant networks. These actions, if effectively implemented, allow optimizing the economic and environmental balance of the management of energy flows within the whole plant.

4 Description of Developed Model Library

The library includes models forecasting off-gases, steam and electricity production and demand by main users. The most representative ML-based ones are here described.

4.1 Off-Gases Production Models

Some of the most important models are related to the production of BFG and BOFG. COG is not predicted, as there are no coke ovens in the considered steelworks. The two models have a common structure; in particular, 3 specialized shallow ESNs are used for predicting respectively gas volume flow, CO and H_2 contents [26, 40–42]. The use of single ESNs for single output is justified by a better accuracy and lower computational cost. The sampling time and prediction frequency is of 5 min for BFG and 1 min for BOFG, as the converter process has a faster dynamic with respect to the BF. The predicted CO and H_2 contents are combined, after some post-processing steps, to obtain the gas NCV, which is combined to the gas volume flow to compute the heating power. For BOFG the model is split in 3 parts: the first one computes the recoverable BOFG, the second one the flareable part of BOFG, the last one the global amount of gas and related energy content. The inputs for the BFG prediction model are the current values of humidity in natural wind, oxygen content in cold wind, cold blast volume flow, pressure of mix wind, steam mass flow in cold blast wind and of amount and main composition of produced BFG. The current values of position of movable skirt, blowed oxygen volume flow and of amount and main composition of produced BOFG are used for the BOFG model together with the values for the 2 h ahead of charged liquid pig iron, charged lime and recycled briquetted sinter. The plant scheduling for the 2 h ahead is a further input in both models in the form of a simple Boolean representation.

4.2 Model of BOF Steam Production

One way to recover the heat of produced off-gases is its transformation in steam. This is the case of steam production during converter operation by exploiting the heat of BOFG. Due to its importance in energy management, a model was developed with a time resolution of 1 min and using a deep layered version of ESN because of the higher accuracy and faster training stage obtained in this situation [26, 43]. The inputs sent to the DESN are the current values of position of movable skirt, blowed oxygen volume flow, recovered BOFG and of the *i-th* converter steam production. Also in this case the Boolean representation of BOF scheduling for the two hours ahead is an input.

4.3 Model of Electrical Power Production by BFG Expansion Turbine

Another way to transform energy between carriers is the expansion of gas to produce electrical power. For instance, BFG after its production is expanded in a dedicated turbine. A model was developed for the prediction of produced electricity [43]. Such model is in cascade to the ones related to the forecasting of BFG amount (see Sect. 4.1). The nonlinear model of the turbine efficiency is realized through a FFNN with a forecasting time frequency of 15 min.

4.4 Model for BFG Demand by Hot Blast Stoves

Hot Blast Stoves (HBS) are important BFG users and are exploited to preheat the air blown into the BF. Two main alternate phases characterize their operation: the heating phase and the wind phase. During the first one, the gas combustion is carried out to heat the walls of the HBSs; this heat is then released to the cold wind during the wind phase. The demand in terms of amount of BFG is forecasted by using an ESN-based model with a time resolution of 5 min [40, 41, 44]. It is fed by the current values of cold blast volume flow, temperature of mix wind and of BFG consumption by each HBS; in addition, Boolean representation of BF plant scheduling and of HBSs operational modes for the 2 h ahead are included in the inputs.

4.5 Model for Steam Injection in BF Cold Blast Wind

One of the use of steam is the injection in BF cold blast wind. A model was developed to forecast this demand, which is based on a standard ESN and is in cascade to the one forecasting the CO and H_2 contents in BFG (see Sect. 4.1), as their values for two hours ahead are some of the inputs. Other inputs are the current values of humidity in natural wind, steam mass flow in cold blast wind, mass flow of pulverized coal, pressure and temperatures of mix wind; a further input is the BF scheduling for 2 h ahead.

4.6 Model for Electricity Demand by BF Fans

BF fans are intensive consumers of electric energy. Thus a model was developed in order to forecast the demand of electrical power by them with a frequency of 15 min. Moreover, this model uses the predicted BFG amount by related model (see Sect. 4.1). Further inputs are the current values of cold blast volume flow, the pressure of mix wind and of the total electrical power supplied to the fans; a further input is the BF scheduling for 2 h ahead.

4.7 Model for Energy Demands by RH Vacuum Degassing

The Ruhrstahl Heraeus Vacuum Degassing (RHVD) process consumes steam and electricity; a simple model for the prediction of these demands was developed by feeding a shallow ESN only with the current values of consumptions and with the Boolean Representation of the scheduling of the plant [43].

5 Results and Discussion

5.1 Prediction of Gas Production and Demand

The obtained accuracy from the models related to the prediction of offgas production and demand are considered suitable for optimization purposes. In particular, the obtained values of the NRMSE are lower than 10% for BFG and lower than 15% for BOFG considering each predicted variable (also intermediate ones), for all forecasting horizon and considering all the possible situations: start-up, stationary and shut down operations. Also in the case of HBSs, the model is capable to follow the process variations and shows a good accuracy with a NRMSE value lower than 12%. Obviously, in all the cases, the NRMSE values increases along the time horizon.

5.2 Prediction of Steam and Electricity Production and Demand

The prediction of production and demand of further energy carriers gives satisfactory results in most of the cases (see Table 1.).

Table 1. NRMSE related to the prediction carried out with the models devoted to steam and electricity production and demand

Model	NRMSE	Note
BOF steam production	$3\% \div 6\%$	
Electrical power production by BFG expansion turbine	$\simeq 0.4\%$	
Steam injection in BF cold blast wind	<20%	For the prediction until 40 min ahead
Electricity demand by BF fans	<5%	
Steam and electricity demands by RHVD	$6\% \div 8\%$ $5\% \div 6\%$	for steam for electricity

Noticeably the model for steam injection in BF cold blast wind shows an accuracy that highly decrease in the prediction time horizon; it is suitable only for the forecasting of 40 min ahead. This can be due either to the fact that the selected inputs are not fully suitable to capture the dynamic of the process or to the predominant role of the operator in decision-making. A refining of the model or of existing procedures are thus required.

5.3 Optimization Results

The hierarchical supervision and control system has been tested through data of a real integrated steelworks. The tests aimed at evaluating the maximum gain which can be obtained through the DSS, thus the supervision and control system was simulated in closed loop, through the implementation of models of the controlled systems, which are characterized by high accuracy. The resulted closed loop system simulation starts from the actual state condition of each plant and exploits data of the real production plan. The results related to two different production period are here reported, showing the performances of the DSS in terms of overall costs sustained for NG and electricity purchase and of amount of gas flared through the torches in comparison to the real case. The first example is related to a period of quite constant production in terms of pig-iron, liquid steel and of refined products, while the second is related to a period of quite constant production of pig-iron and liquid steel but a non-constant production of refined products, namely energy intensive processes were not scheduled. The mentioned processes exploit large amounts of POGs and when they are not scheduled usually the excess gas is flared in the torches.

The results of each period are shown in Fig. 3 and Fig. 4. In particular, Fig. 3.a shows the daily balance of costs related to the controllable external energy sources (NG and electricity) in the top diagram, and the total costs for the overall period (bottom diagram). Figure 3.b shows the energy of the POGs burned in the torches due to their excess in the daily case (top diagram) and the overall POG wasted (bottom diagram).

Figure 4.a and Fig. 4.b show the same indicators related to the second control example. The term *CTRL* is referred to the simulated controlled behavior, while *Real* is related to the actual non-optimized management. All the results are normalized due to confidentiality constraints, with respect to the maximum value of the specific index.

The results of the application of optimization strategy show very encouraging margin of improvement of the POGs network management and in general of the energy media within the steelworks. In particular, the optimized strategy allows rescheduling the POGs flows to the power plant when the internal consumers are not available, by saving huge amounts of NG and by avoiding flaring into torches, with a significant reduction of the environmental impact.

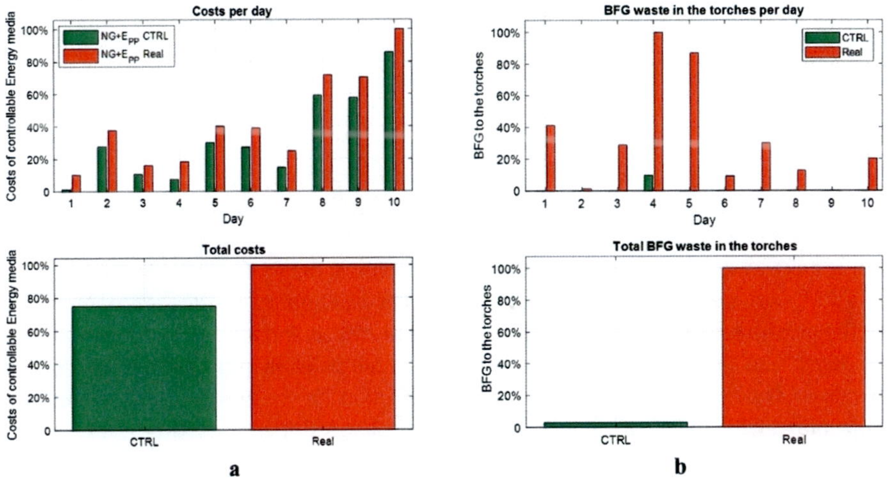

Fig. 3. Results of the optimization strategy for 10 days of quite constant pig iron, steel and of refined products production.

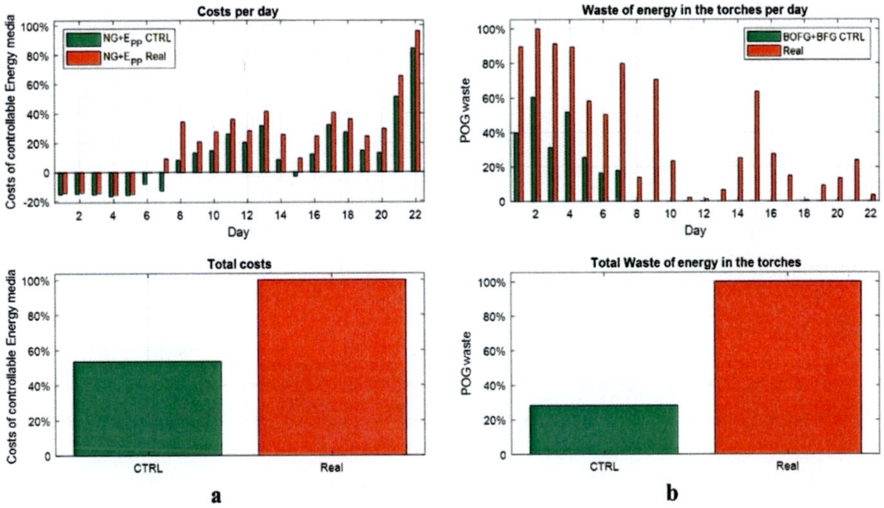

Fig. 4. Results of the optimization strategy for 22 days of non-constant production.

6 Conclusions

The paper provides an overview of a DSS which optimizes the distribution and usage of process off-gases in integrated steelworks, by filling the gaps of standard gas and steam networks management procedures. The core of the DSS are the library of forecasting models and the optimization tool. The models forecast the production and demand of the energy carriers that are involved in the management of the considered distribution networks. ML-based approaches, and in particular ESN, DESN and FFNN, are intensively applied to capture the dynamic of the processes and suitable accuracy for optimization purposes were obtained. The optimization strategy is based on LP and MILP formulations and it is structured as a hierarchical multi-horizon control system and highly benefits from the accuracy and low computational burden of the ML-based models, by achieving very encouraging results. In particular, the hierarchical control system based on EMPC allows reducing the controllable portion of external energy sources between 27.5–46.3% and the exploitation of POG torches between 71.6–96.9% with respect to standard non-optimized management of the POGs flows in the network.

The described application provides a relevant example of application of Artificial Intelligence in order to lower the environmental impact of the steel production processes by also providing a friendly support to the decision processes of the human operators. This shows that the ongoing digitalization process of the steel sector [45], while preserving the centrality of the human role in decision-making, represents a powerful enabler for the improvement of the environmental footprint of the European steel sector and its transition towards an energy-efficient and Carbon-neutral production.

Acknowledgments. The work described in the paper was developed within the project "*Optimization of the management of the process gases network within the integrated steelworks - GASNET*" (Contract No. RFSR-CT-2015-00029) and received funding from the Research Fund for Coal and Steel of the European Union, which is gratefully acknowledged. The sole responsibility of the issues treated in the present paper lies with the authors; the Union is not responsible for any use that may be made of the information contained therein.

References

1. EUROFER, European Steel in Figures (2020). https://www.eurofer.eu/. Accessed 15 Oct 2020
2. EC, What is the European Green Deal?. https://ec.europa.eu. Accessed 15 Oct 2020
3. Janjua, R.: Energy use in steel industry, World Steel Association. http://www.iipinetwork.org. Accessed 15 Oct 2020
4. Steelanol Homepage. http://www.steelanol.eu/en. Accessed 15 Oct 2020
5. Wich, T., Lueke, W., Deerberg, G., Oles, M.: Carbon2Chem-CCU® as a step toward a circular economy. Front. Energy Res. **7**, p. 162 (2020)
6. Deerberg, G., Oles, M., Schlögl, R.: The project Carbon2Chem®. Chem. Ing. Tec. **90**(10), 1365–1368 (2018)
7. Zaccara, A., Petrucciani, A., Matino, I., Branca, T.A., Dettori, S., Iannino, V., Colla, V., Bampaou, M., Panopoulos, K.: Renewable hydrogen production processes for the off-gas valorization in integrated steelworks through hydrogen intensified methane and methanol syntheses. Metals **10**(11), 1535 (2020)
8. Maruoka, N., Akiyama, T.: Exergy recovery from steelmaking off-gas by latent heat storage for methanol production. Energy **31**, 1632–1642 (2006)
9. Kim, D., Han, J.: Techno-economic and climate impact analysis of carbon utilization process for methanol production from blast furnace gas over $Cu/ZnO/Al_2O_3$ catalyst. Energy **198**, 117355 (2020)

10. Gao, R., Zhang, C., Kwak, G., Lee, Y.J., Kang, S.C., Guan, G.: Techno-economic evaluation of methanol production using by-product gases from iron and steel works. Energy Convers. Manag. **213**, 112819 (2020)
11. Shin, S., Lee, J.-K., Lee, I.-B.: Development and techno-economic study of methanol production from coke-oven gas blended with Linz Donawitz gas. Energy **200**, 117506 (2020)
12. Dreillard, M., Broutin, P., Briot, P., Huard, T., Lettat, A.: Application of the DMXTM CO2 capture process in steel industry. Energy Procedia **114**, 2573–2589 (2017)
13. Zhang, Q., Gu, Y.L., Ti, W., Cai, J.J.: Supply and demand forecasting of blast furnace gas based on artificial neural network in iron and steel works. Adv. Mater. Res. **443**, 183–188 (2012)
14. Yang, L., He, K., Zhao, X., Lv, Z.: The prediction for output of blast furnace gas based on genetic algorithm and LSSVM. In: IEEE 9th Conference Industrial Electronics and Applications (ICIEA), pp. 1493–1498 (2014)
15. Zhao, J., Wang, W., Liu, Y., Pedrycz, W.: A two-stage online prediction method for a blast furnace gas system and its application. IEEE Trans. Control Syst. Technol. **19**(3), 507–520 (2011)
16. Zhao, J., Liu, Q., Wang, W., Pedrycz, W., Cong, L.: Hybrid neural prediction and optimized adjustment for coke oven gas system in steel industry. IEEE Trans. Neural Networks Learn. Syst. **23**(3), 439–450 (2012)
17. Liu, Y., Liu, Q., Wang, W., Zhao, J., Leung, H.: Data-driven based model for flow prediction of steam system in steel industry. Inf. Sci. **193**, 104–114 (2012)
18. Porzio, G.F., Fornai, B., Amato, A., Matarese, N., Vannucci, M., Chiappelli, L., Colla, V.: Reducing the energy consumption and CO2 emissions of energy intensive industries through decision support systems–an example of application to the steel industry. Appl. Energy **112**, 818–833 (2013)
19. Porzio, G.F., Colla, V., Matarese, N., Nastasi, G., Branca, T.A., Amato, A., Fornai, B., Vannucci, M., Bergamasco, M.: Process integration in energy and carbon intensive industries: an example of exploitation of optimization techniques and decision support. Appl. Therm. Eng. **70**(2), 1148–1155 (2014)
20. Porzio, G.F., Nastasi, G., Colla, V., Vannucci, M., Branca, T.A.: Comparison of multi-objective optimization techniques applied to off-gas management within an integrated steelwork. Appl. Energy **136**, 1085–1097 (2014)
21. Maddaloni, A., Porzio, G., Nastasi, G., Colla, V., Branca, T.: Multi-objective optimization applied to retrofit analysis: a case study for the iron and steel industry. Appl. Therm. Eng. **91**, 638–646 (2015)
22. Zhao, X., Bai, H., Lu, X., Shi, Q., Han, J.: A MILP model concerning the optimisation of penalty factors for the short-term distribution of byproduct gases produced in the iron and steel making process. Appl. Energy **148**, 142–158 (2015)
23. Kong, H., Qi, E., Li, H., Li, G., Zhang, X.: A MILP model for optimization of by-product gases in the integrated iron and steel plant. Appl. Energy **87**, 2156–2163 (2009)
24. de Oliveira Junior, V.B., Pena, J.C., Salles, J.F.: An improved plant-wide multiperiod optimization model of a byproduct gas supply system in the iron and steel-making process. Appl. Energy **164**, 462–474 (2016)
25. Maddaloni, A., Matino, R., Matino, I., Dettori, S., Zaccara, A., Colla, V.: A quadratic programming model for the optimization of off-gas networks in integrated steelworks. Matér. Tech. 107(5), p. 502 (2019)
26. Colla, V., Matino, I., Dettori, S., Cateni, S., Matino, R.: Reservoir computing approaches applied to energy Management in industry. Commun. Comput. Inf. Sci. **1000**, 66–79 (2019)

27. Cateni, S., Colla, V., Nastasi, G.: A multivariate fuzzy system applied for outliers detection. J. Intell. Fuzzy Syst. **24**(4), 889–903 (2013)
28. Cateni, S., Colla, V., Vannucci, M.: A fuzzy logic-based method for outliers detection. In: Proceedings of the IASTED International Conference on Artificial Intelligence and Applications, AIA 2007, pp. 561–566 (2007)
29. Cateni, S., Colla, V., Vannucci, M.: General purpose input variables extraction: a genetic algorithm based procedure GIVE a GAP. In: 9th International Conference on Intelligent Systems Design and Applications, ISDA 2009, pp. 1278–1283 (2009)
30. Cateni, S., Colla, V., Vannucci, M.: A genetic algorithm-based approach for selecting input variables and setting relevant network parameters of a SOM-based classifier. Int. J. Simul. Syst. Sci. Technol. **12**(2), 30–37 (2011)
31. Vannucci, M., Porzio, G.F., Colla, V., Fornai, B.: Use of clustering and interpolation techniques for the time-efficient simulation of complex models within optimization tasks. In: 2011 UKSim 5th European Symposium on Computer Modelling and Simulation, IEEE, pp. 6–11 (2011)
32. Jaeger, H.: The "echo state" approach to analysing and training recurrent neural networks-with an erratum note. Bonn, Ger. Ger. Natl. Res. Center Inf. Technol. GMD Tech. Rep. **148**(34), 13 (2001)
33. Jaeger, H., Haas, H.: Harnessing nonlinearity: predicting chaotic systems and saving energy in wireless communication. Science **304**(5667), 78–80 (2004)
34. Grigoryeva, L., Ortega, J.P.: Echo state network are universal. Neural Networks **108**, 495–508 (2018)
35. Deihimi, A., Showkati, H.: Application of echo state networks in short-term electric load forecasting. Energy **39**(1), 327–340 (2012)
36. Shi, G., Liu, D., Wei, Q.: Energy consumption prediction of office buildings based on echo state networks. Neurocomputing **216**, 478–488 (2016)
37. Gallicchio, C., Micheli, A.: Echo state property of deep reservoir computing networks. Cogn. Comput. **9**(3), 87–99 (2017)
38. Gallicchio, C., Micheli, A., Pedrelli, L.: Design of deep echo state networks. Neural Networks **108**, 33–47 (2018)
39. Wolff, A., Mintus, F., Bialek S., Dettori, S., Colla V.: Economical mixed-integer model predictive controller for optimizing the sub-network of the BOF gas, METEC 4th ESTAD, Düsseldorf (2019)
40. Dettori, S., Matino, I., Colla, V., Speets, R.: Deep echo state networks in industrial applications. In: IFIP Advances in Information and Communication Technology, 584 IFIP, pp. 53–63 (2020)
41. Matino, I., Dettori, S., Colla, V., Weber, V., Salame, S.: Forecasting blast furnace gas production and demand through echo state neural network-based models: pave the way to off-gas optimized management. Appl. Energy **253**, 113578 (2019)
42. Matino, I., Dettori, S., Colla, V., Weber, V., Salame, S.: Application of echo state neural networks to forecast blast furnace gas production: pave the way to off-gas optimized management. Energy Procedia **158**, 4037–4042 (2019)
43. Dettori, S., Matino, I., Colla, V., Weber, V., Salame, S.: Neural Network-based modeling methodologies for energy transformation equipment in integrated steelworks processes. Energy Procedia **158**, 4061–4066 (2019)
44. Matino, I., Dettori, S., Colla, V., Weber, V., Salame, S.: Two innovative modelling approaches in order to forecast consumption of blast furnace gas by hot blast stoves. Energy Procedia **158**, 4043–4048 (2019)
45. Branca, T.A., Fornai, B., Colla, V., Murri, M.M., Streppa, E., Schröder, A.J.: The challenge of digitalization in the steel sector. Metals **10**(2), 288 (2020)

Industrial Cyber Security at the Network Edge: The BRAINE Project Approach

Francesco Paolucci[1], Pavel Burget[2], Victor Genin[3], Ela Kolis[3], Gheorghe Dobra[4], Lucian Cotet[4], Irina Busila[4], Sylvain Guilley[5], Philippe Nguyen[5], Youssef Souissi[5], Liran Liss[6], Boris Pismeny[6], Juan Jose Vegas Olmos[6], and Filippo Cugini[1](✉)

[1] CNIT, Pisa, Italy
{francesco.paolucci, filippo.cugini}@cnit.it
[2] Czech Technical University, Prague, Czech Republic
pavel.burget@cvut.cz
[3] Librimind, Jabotinsky 33, 5251107 Ramat-Gan, Israel
{victor, ela}@librimind.com
[4] Alum, 820228 Tulcea, Romania
dobra@alro.ro, {lcotet, ibusila}@alum.ro
[5] Secure-IC S.A.S., Tour Montparnasse, 27th floor, 75015 Paris, France
{sylvain.guilley, philippe.nguyen, youssef.souissi}@secure-ic.com
[6] NVIDIA, Hakidma 26, 2069200 Yokneam, Israel
{liranl, borisp, juanj}@nvidia.com

Abstract. This work presents the BRAINE Project approaches and solutions to enable AI-empowered industrial applications to efficiently and securely operate at the network edge. First, two use cases are presented to highlight the motivations for secure edge computing in industrial plants, including low-latency applications and the need to locally handle production data in a secure way. Then, AI-based proactive and reactive approaches are discussed to provide comprehensive cyber security at the industrial edge. Finally, two specific hardware accelerated implementations for cyber security are presented. The first implementation focuses on data encryption at network interface cards, the second implementation provides deep packet inspection for reconnaissance detection in switching elements. Results that hardware accelerated solutions enable the effective deployment of security solutions guaranteeing wire-speed operations while freeing CPU resources for industrial applications.

Keywords: Cyber security · Edge computing · Artificial Intelligence · AI · Machine Learning · ML · Network interface cards · NIC · P4

1 Introduction

Manufacturing in Europe represents approximately 21% of its GDP and 20% of its employment, providing more than 30 million jobs in over 230,000 enterprises. This sector has a key role in the European Union's (EU) economic growth and social development. New paradigms are emerging within the manufacturing ecosystems:

V. Colla and C. Pietrosanti (Eds.): ESTEP 2020, AISC 1338, pp. 119–131, 2021.
https://doi.org/10.1007/978-3-030-69367-1_10

predictive maintenance, in-factory optimization, human-robot collaboration (HRC), and digital twinning are now considered evolutionary steps. Combining machine-to-machine communication with big data analytics in industry can increase efficiency, productivity, and performance. And the adaption of those paradigms is a matter of sustainability over the long run as well as global competitiveness. However, some of these paradigms are taking time to be absorbed, partly because the security concerns associated with introduction of digital technologies and cyber-attacks create a reluctance in industry for a wider adoption without isolation and proprietary solutions. According to Gartner, there are currently over 6 billion IoT devices on the planet. Such a huge number of potentially vulnerable elements may be exploited by cyber criminals. Nowadays, several thousand different malware have been discovered for IoT devices, and the number is growing by twice each year. The IoT devices are often vulnerable to attacks because of the high number, limited processing capabilities and low manageability, which makes it hard to keep all the devices updated and secured.

To decrease the risk of attack and minimize the influence of cyber vulnerabilities on the devices a comprehensive solution is needed, covering computing, networking, industrial applications, and data management [1, 2]. In addition, secure industrial solutions have to account for the recent evolutions of production processes, driven by Artificial Intelligence (AI), which often require low latency performance. This is driving the introduction of AI-empowered secure edge computing solutions within industrial plants [3–5].

In this paper we present a subset of the activities carried out within the BRAINE Project [6], a use-case-driven EU-funded initiative targeting the development of Edge Micro Data Center solution for secure, AI-empowered industrial edge computing.

2 Use Cases

2.1 Securing Edge-Based Solutions for the Aluminium Industry

ALUM SA is one of the largest producer in Central and Eastern Europe of calcined alumina and aluminium hydroxide in Romania. ALUM Tulcea refinery has been commissioned in 1973, initial designing capacity was 250,000 mt/y and through several expansions the production nominal capacity was increased to 600,000 mt/y. The company recently has endowed the Research and Development Department with new equipment in order to efficientize the existing research facilities for supporting the economic competitiveness and business development and also to be in compliance with the European norms on environment protection (Fig. 1).

With recent EU emission targets (COM/2019/640, The European Green Deal), there is a great focus to make the production process more efficient. Alum has identified several potential processes for optimization. One of them is the comminution (grinding) of aluminium hydroxide. The Process Industry heavily relies on grinding which consumes up to 4% electrical energy globally [7] and the highest consumer of energy. The grinding process involves the dry aluminium hydroxide resulting from the drying stage, which is prior to the grinding stage and is the feeding the grinding equipment. The grinding equipment has ceramic shields and grinding media (Al2O3). The equipment operation is

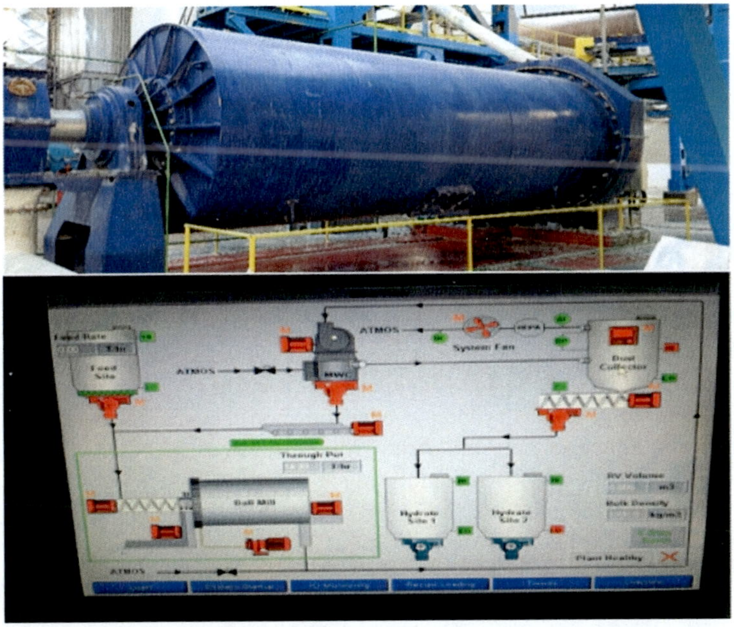

Fig. 1. Production environment and its control system

made by the measuring and control loops, thus providing a high operation accuracy, and all the operation data, the equipment setup parameters (minimum and maximum levels of dry hydrate in the feeding silo, minimum and maximum levels of dry and grinded hydrate in the finished product storage silo, dry hydrate equipment supply flowrates, number of rotations per minute on the grinding equipment, bearing temperatures, bag filter pressure, dust emissions after the dedusting operation etc.) are monitored and processed by the research department staff. A reduction of process costs in the comminution correlated with advanced processes through digitalization directly reduces energy consumption and relate emissions levels, which would benefit the company by reducing the costs as well as the future carbon footprint to be aligned with new government and EU emission targets. However, introducing digitalization and cloud-based solutions into the production process poses a serious challenge from the cyber-security and data management perspective. Historically, only the production staff had access to the production data, however with the integration of digital optimization tools, more people, including outside experts, are able to access the production data for suggested improvements. In addition, offloading the computation to the cloud introduces additional risks on data management. New approaches based on edge computing represent the most suitable solution for digitalization since they enable low-latency AI applications for process optimization while guaranteeing adequate cyber security control with full data ownership. BRAINE aims to develop a comprehensive security platform ranging from computing to networking solutions, statistical models of inference and machine-learning algorithms to learn how to imitate the desired behavior of targeted devices and generate statistically similar data associated with the sensor devices.

2.2 Securing Digital Twin for Industry 4.0 at the Edge

Demand for lower time to market period, product customization, and production in small lots brings a lot of focus on flexibility, reconfigurability, and production scalability. In classical production, monolithic Manufacturing Execution Systems (MES) and Enterprise Resource Planning (ERP) are trying to fulfill these requirements. In our use case, we are utilizing the new strategy to MES level management and planning using a multiagent approach to better comply with those requirements. This approach uses agents as self-sufficient units to negotiate the desired behavior on the go and enables the use of distributed algorithms for production planning. Multiagent planning allows us to handle more computationally complex tasks than in the case of centralized planning, and at the same time, agents itself can be migrated and duplicated among the edge devices to balance the load and lower the response time dynamically. For the system supervision, a domain agnostic motif/pattern discovery tool, which operates on multidimensional time series (MOD) will be used. This tool can detect anomalies in production processes and provide inputs for association analysis of the device's parameters and products' parameters. This tool will be modified to run on the edge and feed the Digital Twin in the Cloud. The whole multi-agent architecture is depicted in Fig. 2. The entire architecture consists of four layers. The Hardware layer

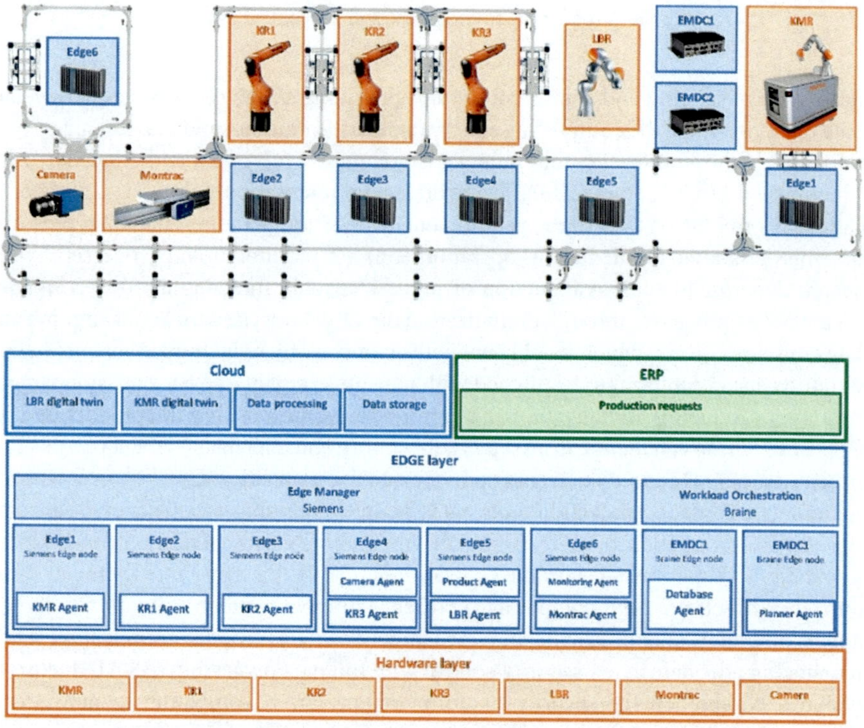

Fig. 2. The upper part shows the Testbeds production line with devices corresponding to the blocks in the diagram at the bottom. The bottom portion depicts the architecture of the multi-agent system itself.

represents devices directly influencing the manufacturing process on the shop floor. These devices contain only simple methods and communicate directly with their agents to acquire complex decisioning. Examples of those devices are robots, sensors, or conveyors. The Edge layer consists of computing units used as the agent runtime environment. Each device in this layer supports Docker containers and is orchestrated by the Kubernetes management system. Edge layer devices are on one side connected directly to devices in the hardware layer and on the other connected to Cloud and ERP. The Cloud is used for storing and processing data and storing Digital Twins. The ERP block is used for business process management software that allows an organization to use a system of integrated applications to manage the business.

The use case will be implemented on the production line in CTU's Testbed for Industry 4.0. For this purpose, will be used Siemens Edge devices together with BRAINE Edge Micro Data Center devices connected into one managed unit utilizing Kubernetes.

3 BRAINE's Approach

3.1 Overall Architecture

BRAINE [6] is a project that aims at realizing a full-stack solution based on the state-of-the-art and beyond knowledge and technology in the areas such as Low Energy Hardware, Edge Computing, AI. The solution is packaged as a single-box Edge Micro Data Center (EMDC) containing all the required hardware and software needed for ultra-high-performance applications such as human organ digital twins for healthcare assistance, large scale video surveillance for smart city scenarios, and autonomous industrial management and control systems.

In this subsection, we briefly introduce the high-level architectural design of the BRAINE system.

As shown in Fig. 3, the BRAINE approach leverages on a five layer architecture. Layer 1 (L1) enables the EMDC to communicate with external systems such as cloud services, legacy and enterprise systems, IoT systems and other EMDCs.

Layer 2 (L2) builds a cluster-in-the-box with a set of worker nodes that are managed by a management node. The worker nodes utilize various processing units e.g., CPUs, GPUs, and FPGAs, different amounts of memory and storage. The workers' communication is managed based on shared-nothing-architecture, hence only through the network.

Layer 3 (L3) manages the cluster via a customized Kubernetes, in that the security requirements and privacy policies are taken into account in workload placement as well as data placement. The customized Kubernetes scheduler also considers Service Level Agreements (SLAs) such as latencies and response time that are important for industrial applications. Layer 3, also provides distributed file system over the whole cluster to provide a higher abstraction layer for the application. All the resources and services available to the cluster manager are registered a knowledge graph which is utilized by other components such as the workflow definition tool.

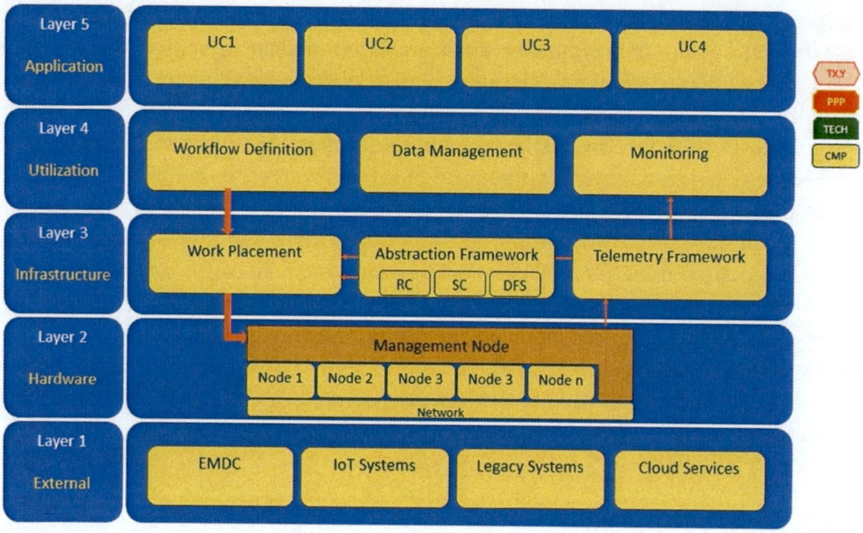

Fig. 3. BRAINE Architecture.

Layer 4 (L4) focuses on workflow definition, data management and infrastructure monitoring for application service level verification. Layer 5 (L5) includes the applications for the use cases.

In the BRAINE project, secure technologies are implemented across all layers, providing a comprehensive secure solution that ranges from hardware to software up to the use case applications and their data management, specifically designed for industrial systems.

4 BRAINE Cyber Security Approach

4.1 Security Threat Model

Industrial systems face multiple threats. Indeed, the facilities (called "plants" in the sequel) are complex hence offer a large attack surface. Moreover, plants are valuable profit centers and thus represent natural targets.

Understanding threats is key in building up a protection which is in relationship with the expected damages. One method consists in analyzing threat agents. They can be classified very versatilely, such as:

- Individuals with limited information and limited budget
- Founded opponents or insiders
- Motivated agencies

Another view on the threat is the expression of the expected goal:

- block the plant and request for money to unblock (ransomware)
- sabotage in order to defeat competition (malware)
- steal information about industrial processes, clients' identities and customizations, etc. (spyware)

Still another view is through the prism of identification of attack paths:

- network ports
- air gap (banal) USB stick purportedly left inadvertently and unattended near the premises (albeit containing a real backdoor)
- rogue maintenance and/or housekeeping personnel, insider, or former/fired employee.

It can be clearly seen that the asymmetry between the attacker and the defender of the plant is huge: the plant shall protect against a countless amount of attacks while the attacker shall only succeed a single one.

For this reason, good practices are preferred, since they allow to cover most of above mention threats in a preventive manner. This is the idea behind IEC 62443, which is a methodological guide for secure implementation.

It shall be noted that some current markets opted for a different approach. For example, automotive market is completely risk-analysis based. A methodology to use risks as input and derive a test plan is typically proposed in Singaporean draft TR68-3.

This approach is captured by the solution we later describe: it consists in ensuring a detection of any deviation, whoever, for whichever reason, wherever the attack originates. It is based on the (machine) learning of a nominal situation, and flags as safety or security issues any deviations.

4.2 Proactive/Reactive Cyber Security Methodologies Based on AI

It is obvious that when the system targeted by an attacker is defenseless, the later will be lazy and apply plain state-of-the-art attacks. But when there are defense mechanisms in place, the challenge is larger for the attacker, and he also parallely gets excited about the perspective of overcoming the measures put in place against him.

In the field of cyber-security, the winning mindset is to consider that attacks are adaptive and clever adversaries which in addition think out of the box. Thus, innate nature of the cyber-threat is to remain essentially surprising. Therefore, solutions which ensure all the same a large coverage shall be versatile. This is where machine learning comes into play. The solution consists in two stages: information collection and information analysis.

The first aspect concerns the proactive deployment of hardware sensors and software probes. Indeed, attacks can only be detected provided a swarm of various "antennae" are deployed. Advantageously, the sensors are heterogeneous, i.e., they collect different kinds of information. Such setup maximizes the device awareness with respect to domain of possible malevolent actions. Indeed, those might be diverse, ranging from:

- Initial firmware infection or configuration manipulation,
- Followed by silent wait for remote control & command signals,
- Until the execution proper, i.e. the payload delivery (which can consist in various actions, such as data exfiltration, system compromising, denial of service, etc.).

The second aspect is the reactive analysis of the measurements. This component in the protection system consists in implementing the "smart" layer over the low-level "antenna". Such a supervision by machine learning is all the more relevant as, moving forward, edge hardware is being more and more empowered by AI capability. Therefore, the attack detection can be carried out *in situ*, for faster reaction and also to withstand network bottlenecks between the plant and the SOC (Security Operation Center) that would otherwise be created by the large flux of data to upload. Secure-IC deploys the "Securyzr Server" technology to meet these requirements in terms of bandwidth [8]. Learning can be performed according to various strategies. When the feared event is well characterized, then some jig can be resorted to clearly separate admissible conditions from abnormal conditions. But this situation is exceptional. Most of the time, the system designer has only a vague idea of the attack effects. Hence learning is based on two scenarios:

1. First of all, in any nominal condition, taking into consideration all the benign environmental variability (temperature/power/humidity range, affluence of operators in facilities, etc.)
2. Second, in emblematic threat conditions, which ought to be simulated (actually placing the plant in danger for the sake of learning its behavior can turn out to be an accidental attack).

This allows for binary classification from the two abovementioned situations.

But there is also the possibility that the plant designer merely cannot anticipate the outcome of a protean attack. The detection must thus be extremely flexible, and recommended approaches are termed "single class" or "outlier detection". Machine learning algorithms are typically "One-Class Support Vector Machine" (OCSVM), "Elliptical Envelope" (EE), "Isolation Forest" (IF), or "Local Outlier Factor" (LOF).

Still, it is important to diagnose attacks, in a view to perform forensic analysis and also to better understand the attacker so as to stop his actions dead in the future. Hence the outcome of the reactive part can be refined (compared to a sheer attack/no-attack binary classification). One example is sketched in Fig. 4. This figure provides a decision diagram for hardware attacks. Of course, it requires an attack ontology to label the various options of attack characterization.

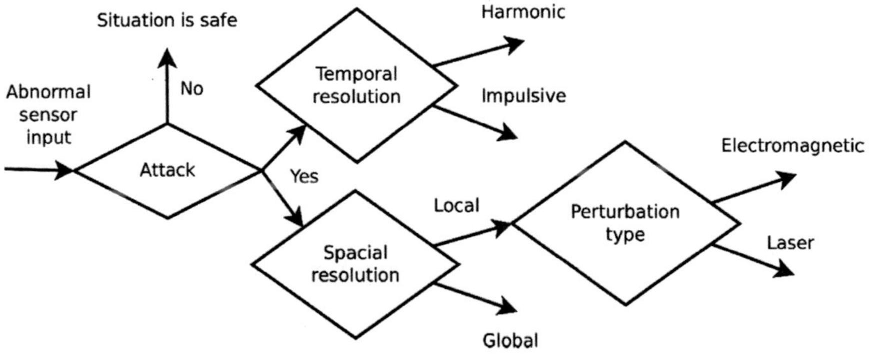

Fig. 4. Binary decision diagram to refine the characterization of an attack, illustrated on the example of a hardware perturbation.

5 Specific BRAINE Solutions for Industrial Cyber Security at the Edge

Based on the aforementioned security approach and methodologies, in this section we provide two specific security solutions leveraging on hardware acceleration targeting (1) secure encrypted communication and (2) cyber-attack detection and mitigation.

5.1 HW Acceleration for AI and Secure Communications

Edge computing platforms enabling AI in factory scenarios can only be successful if they comply with a series of requisites: ease of use, scalable, secure and trustworthy. In terms of software, the trend by providers is to rely on open software libraries and software development kits (SDKs) which allow customers to deploy inference and AI in the cloud, on their servers, or at the edge. This software investment is designed to accelerate customer time-to-deployment of and reduce the overall development costs.

Contrary to the software investment, which mainly takes into account development costs, hardware investments are considered with a wide range of parameters in mind: from space front print, cost of maintenance, scalability, latency requirements, upgradeability and performance, among others. Fundamentally, hardware acceleration or hardware offloading, resolves many of these aspects, and in co-existence with the networking aspects of the edge platform, also improves the security grading of the infrastructure.

Hardware acceleration is therefore a critical aspect affecting primarily networking operations, either when shuffling data around for processing or storage, which are in its nature inherently a must for AI operations. To secure edge node connectivity at low latency and ensure AI performance is not impacted, there is a wealth of work on

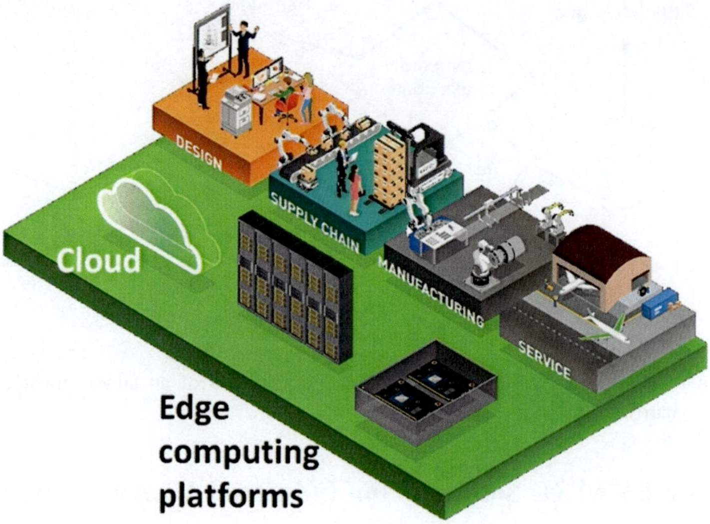

Fig. 5. Factories relying on edge computing platforms.

hardware inline encryption IPsec and TLS; particularly, AES-GCM 128/256 (effectively the gold standard for cryptography, supporting 128/256 bits), IPsec encapsulation/decapsulation and Extended Sequence Numbers (ESN). Crypto-acceleration brings to the edge a full securitization of the communications at the edge, while drastically reducing the added latency and freeing up CPU resources, which can be used for AI processes instead. The following Fig. 6 shows different simulation results for regular https traffics vs. HW offloaded traffic for 1 or 8 cores for offloading at the network interface cards interfacing the processing blocks at the edge (the horizontal axis is the size of the data being transferred).

In Fig. 6, "https" represents the performance of encrypted communication with no HW acceleration while "http" represents the upper bound, where no encryption is applied. "Offload" and "offload+zc" are two proposed offloading schemes leveraging on HW acceleration. Figure 6(left) shows that a single core can reach only extremely low throughput. Figure 6(center) shows that 8 cores are needed to achieve the maximum throughput of 100 Gb/s supported by the considered interface. Compared to "https" performance, the two HW-based solutions significantly increase the transfer rate, achieving performance rather close to the "http" bound. Figure 6(right) shows the busy cores. As it can be observed, for large data transfers, by all metrics the best performance is achieved for offloaded.

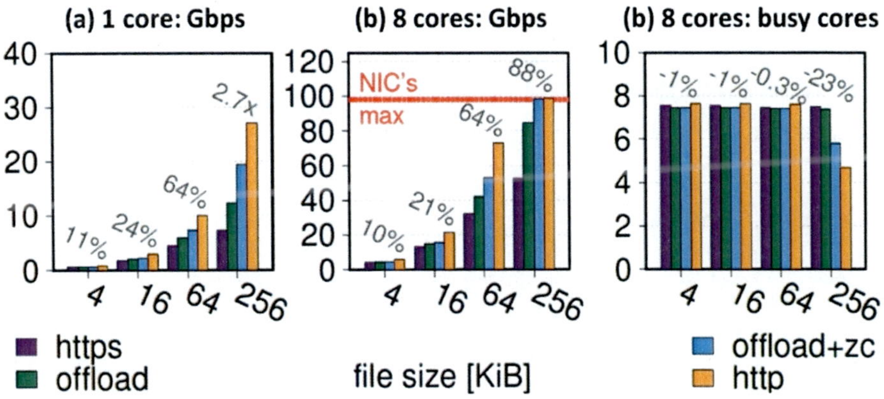

Fig. 6. Scalability of HW acceleration for secure communications.

5.2 Programmable Networking for Industrial Cyber Security at the Edge

The potential of the previously described HW acceleration technology is further increased by its capability to enable customized network offloading operations at wire speed, leveraging on P4-based programmability. P4 is a high-level data plane programming language enabling the flexible definition of a custom pipeline (sequence of match-action tables), configurable control workflows, metadata elaboration, and stateful capabilities. These features allow complex packet manipulation and flexible processing procedures that are currently not available in standard switches. Novel dynamic traffic engineering solutions can be defined at the edge to dynamically modify forwarding actions accounting for traffic rates, to account for flow matches occurred during a specific time interval, or to preserve time sensitive performance. In addition, novel cyber security solutions can be embedded in any network device, accelerating AI-based threat detection while constituting a distributed barrier against cyber-attacks.

Security breaches affecting sensors or industrial IoT devices can severely compromise the entire industrial network. Reconnaissance and Denial of Service mitigation inside an industrial networking element can be programmed with a P4 code exploiting stateful objects. With respect to stateless firewalls, the P4 processing is able to store and correlate protocol field values belonging to correlated sessions, thus identifying suspected flows (e.g., port scan behavior) that may be dynamically blocked.

Figure 7 shows the P4-NetFPGA latency performance (10 Gigabit Ethernet interfaces) in the case a cyber-attack occurs while 10.000 entries are present in the flow tables [9]. The port scan attack is successfully detected and blocked without impacting the overall performance. Results show an average latency of 5 μs, with very low variance (i.e., min and max latencies differ of around 50 ns), independent from the traffic throughput in the range of 0–9.6 Gb/s. These results are substantially the same with or without the extra P4 code needed for attack detection and mitigation, proving that the P4 processing for cyber-security can be enforced without introducing significant latency degradation.

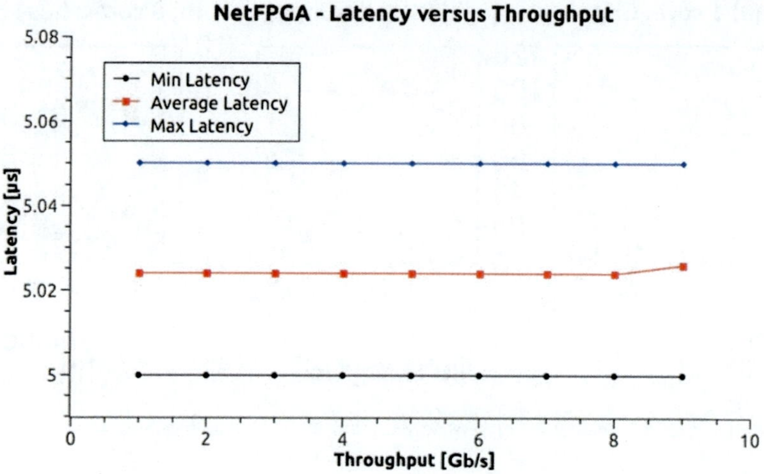

Fig. 7. Scalability of in-network DoS attack detection.

6 Concluding Remarks

In the steel industry, edge computing is emerging as a suitable solution to enable effective AI-based process optimization of manufacturing systems and production processes. In addition, edge computing enables full ownership of critical production data. However, to take full benefit of edge computing, adequate security solutions need to be implemented. Indeed, cyber-attacks could determine an enormous impact, potentially affecting the working operations of machinery and production processes, possibly degrading the integrity of the steel, causing massive physical damage when critical process components become unregulated, and damaging the relationships with customers or investors.

This work presented a selection of activities carried out in the BRAINE Project, which aims at boosting big data processing and artificial intelligence at the network edge. The Project specifically targets industrial scenarios where low latency applications need to be implemented using powerful hardware infrastructures, flexible software platforms and efficient data management, all supported by comprehensive security solutions. This paper presented the motivations for secure edge computing in industrial environments by introducing two use cases, the first on aluminium production, the second on digital twin for industry 4.0. The BRAINE approach for cyber security is then introduced, showing the proactive/reactive cyber security methodologies based on AI. Furthermore, two examples of BRAINE implementations for cyber security leveraging on hardware acceleration are presented. Results showed that hardware accelerated solutions are capable of operating at wire-speed and free CPU resources which can be dedicated to industrial applications and optimization of steel production processes.

Acknowledgement. This work has received funding from the BRAINE Project ("Big data pRocessing and Artificial Intelligence at the Network Edge"), ECSEL Joint Undertaking (JU) under grant agreement No 876967. The JU receives support from the European Union's Horizon 2020 research and innovation programme and Italy, Poland, Denmark, Netherlands, Ireland, Hungary, Germany, France, Slovakia, Bulgaria, Finland, Czech Republic, Israel, Switzerland.

This study was possible by implementation of the "Endow the Research and Development Department of SC ALUM SA Tulcea with independent and efficient research facilities to support the economic competitiveness and business development" project, co-funded by the European Regional Development Fund through the Competitiveness Operational Programme 2014–2020. Under this project were purchased and commissioned: "Independent equipment/Installation for research and development of the technology of wet aluminum hydroxide classification", "Independent equipment/Installation for research and development of technology to obtain the dried aluminium hydroxide" and "Independent equipment/Installation for research and development of the technology of grinding and screening the dried aluminum hydroxide".

References

1. Dai, W., et al.: Industrial edge computing: enabling embedded intelligence. IEEE Ind. Electron. Mag. **13**(4), 48–56 (2019)
2. Wang, T., et al.: A unified trustworthy environment establishment based on edge computing in industrial IoT. IEEE Transa. Ind. Inform. **16**(9), 6083–6091 (2019)
3. Kaur, K., et al.: Edge computing in the industrial internet of things environment: software-defined-networks-based edge-cloud interplay. IEEE Commun. Mag. **56**(2), 44–51 (2018)
4. Qiu, T., et al.: Edge computing in industrial internet of things: architecture, advances and challenges. IEEE Commun. Surv. Tutorials **22**(4), 2462–2488 (2020)
5. He, D., Chan, S., Guizani, M.: Security in the Internet of Things supported by mobile edge computing. IEEE Commun. Mag. **56**(8), 56–61 (2018)
6. BRAINE Project, "Big data pRocessing and Artificial Intelligence at the Network Edge", ECSEL Joint Undertaking (JU), grant agreement No 876967. https://www.braine-project.eu
7. Jeswiet, J., Szekeres, A.: Energy consumption in mining comminution. Procedia CIRP, **48**, 140–145 (2016)
8. https://www.secure-ic.com/solutions/securyzr/
9. Paolucci, F., Civerchia, F., Sgambelluri, A., Giorgetti, A., Cugini, F., Castoldi, P.: P4 edge node enabling stateful traffic engineering and cyber security. J. Opt. Commun. Networking **11** (1), A84–A95 (2019)

Smart Steel Pipe Production Plant via Cognitive Digital Twins: A Case Study on Digitalization of Spiral Welded Pipe Machinery

Özlem Albayrak(✉) and Perin Ünal

Teknopar Industrial Automation,
İvedik OSB 1471. Street no: 3-5, 06370 Yenimahalle/Ankara, Turkey
{albayrak, punal}@teknopar.com.tr

Abstract. There are many Artificial Intelligence (AI), Machine Learning (ML) and Deep Learning (DL) applications being used in real manufacturing plants. Digital twins with AI/ML/DL capabilities are among these applications. Steel is an important process industry, where such applications have vast potential to provide positive impacts on different aspects, including energy consumption and cost benefits. The Cognitive Plants Through Proactive Self-Learning Hybrid Digital Twins (CogniTwin) project aims at adding the cognitive elements to the existing process control systems, enabling their capability to self-organise and offer solutions to unpredicted behaviours. In NOKSEL pilot of CogniTwin, the main problem is related to very high costs of machines breakdowns. By developing digital, hybrid and cognitive twins, the pilot aims at reducing energy consumption and average duration of machine downtimes. Cognitive twins for Spiral Welded Steel Pipes (SWP) will enable predictive maintenance at the steel pipe plant. This study briefs the NOKSEL pilot's purpose, scope, state and the results gained in the 1st year of the project.

Keywords: Digital twin · Cognitive twin · Industry 4.0 · Spiral welded steel pipe · Artificial intelligence · Predictive maintenance · Smart condition monitoring

1 Introduction

Recent developments in AI and Digital Twins offer new possibilities that have enormous potentials to redefine the existing and future of manufacturing [19, 22]. As a result, Digital Twins, together with AI, ML and DL technologies are gaining ever increasing attention both in academia and in various industries. Providing great capabilities in order to realize Industry 4.0, Digital Twins enhanced with AI technologies not only enable real-time working conditions simulations, but also provide intelligent decision-makings related to important parameters in manufacturing [8, 16, 20, 26].

Cognitive Digital Twinning is an important trend in many applications, and the steel production industry is not an exception. Similar to the other industries, the steel

V. Colla and C. Pietrosanti (Eds.): ESTEP 2020, AISC 1338, pp. 132–143, 2021.
https://doi.org/10.1007/978-3-030-69367-1_11

industry has also been facing rapid and constant changes due to Industry 4.0 revolution. With about 500 active production centers in 24 Member States of European Union, the steel industry represents an important asset for the European economy. Europe is the largest importer of steel and the second largest producer of steel in the world after China [12].

By 2024, condition monitoring market is expected to witness high growth value like 3.5 billion USD [12]. According to European Steel Skills Agenda (ESSA), Digital Twin related sales is estimated to reach up to 18.29 billion USD in 2024 [12].

The remaining of the study is organized as follows: In Sect. 2, following presentations of Digital Twin and Cognitive Digital Twin concepts background information, Digital Twins and AI/ML and DL technologies utilizations in the steel industry are given. Section 3 briefs the CogniTwin project. Section 4 details the NOKSEL case by providing the case objectives, success criteria, work accomplished during the first year of the CogniTwin project in the NOKSEL case, and remaining planned future work for the case. Finally, Sect. 5 concludes the study.

2 Background

As reported by Grand View Research, Inc., by 2025 the global market of Digital Twin is expected to reach up to 26.07 billion USD, with a Compound Annual Growth Rate of 38.2% [22]. Awareness on Digital Twin technology as an enabling tool for bridging physical and cyber world has been growing exponential [24]. Due to important impacts of Digital Twins, several surveys and reviews studying different aspects of Digital Twins and their applications have been conducted [5, 15, 23, 25].

2.1 Digital Twins and Cognitive Digital Twins

There are multiple definitions of Digital Twin proposed by different organizations and studies. Although there are various definitions exist, the core of Digital Twin remains the same [24]. Some of the definitions for Digital Twin are given in the following sentences:

"The Digital Twin is the effortless integration of data between a physical and virtual machine in either direction" [15].

"Digital Twins are the virtual representations of resources organizing and managing information and being tightly integrated with AI, ML and cognitive services to further optimize and automate production" [18].

"Digital Twin is a digital duplication of entities with real-time two-way communication enabled between the cyber and physical spaces" [I-Scoop 2017, in 24].

"A Digital Twin can be defined as an adaptive model of a complex physical system" [29].

Table 1 in [24] list other definitions of Digital Twin made in previous studies.

According to Industrial Internet Consortium (IIC), a Digital Twin is defined as a formal digital representation of some asset, process or system that captures attributes and behaviors of that entity suitable for communication, storage, interpretation or processing within a certain context [10]. The definition provided by Abburu et al. is

Table 1. CogniTwin participants

Industrial partners	Universities and R&D institutes	Technology providers
• Noksel Steel Pipe • Hdyro Aumınium Deutschland GmbH • SHI FW Energia Oy • Sidenor Aceros Especiales Europa SL • Elkem ASA • Saarstahl AG	• SINTEF AS • The German Research Center for Artificial Intelligence (DFKI) • Fraunhofer-Gesellschaft • University of Oulu	• Teknopar Industrial Automation • Cybernetica AS • Nissatech Innovation Centre • Scortex

parallel to IIC's description, where a Digital Twin is defined as "a digital replica of a physical system that captures attributes and behaviors of that system" [1].

Based on the relations among digital twins in systems, IIC introduces five different categories of digital twins: 1) discrete, 2) composite, 3) hierarchical, 4) associational and 5) peer-to-peer [10].

Digital Twins can serve as the basis for advanced analytics and AI applications. Plus, advanced analytics and AI applications can be part of a Digital Twin, making it an intelligent and self-contained entity [10]. Abburu et al. defines three types of digital twins: 1) digital twins, 2) hybrid digital twins, and 3) cognitive digital twins. A hybrid twin is a digital twin, and a cognitive twin is a hybrid twin (Fig. 3). A digital twin is a physical replica of a physical system. A hybrid twin using different set of data (such as sensor data, databases, simulators, etc.), contains a set of interconnected models, where a cognitive twin is proactive and learn by itself. The authors state that the "cognitive element" in the Digital Twins will be introduced by learning from historical process data and events to predict unwanted events in the operation before these events happen (Fig. 3) [1].

Cognitive digital twins are special types of digital twins that include cognitive elements, and not every Digital Twin is a Cognitive Digital Twin. The research trends related to Digital Twins and Cognitive Digital Twins (i.e. Digital Twins including cognitive elements due to utilization of AI, ML and/or DL) seem to follow parallel trends (Fig. 1 and Fig. 2).

Our Google Scholar search, conducted on 2020.11.09 using the search query: "digital twin" AND ("artificial intelligence" OR "machine learning" OR "deep learning" OR "cognitive"), returned 28 results between the years 2013–2014, and 5430 between 2019–2020 (Fig. 1). When searched only in the Title of the publications, the same query, returned zero and 15 studies respectively (Fig. 1).

Some of the studies conducted on digital twins exclude the terms: "artificial intelligence", "machine learning", "deep learning", and "cognitive", hence these studies only focus on Digital Twins that are not cognitive, or that do not contain AI elements.

Despite the separation of Digital Twins and Cognitive Digital Twins, some authors consider all Digital Twins to be supported by AI [16, 21]. Eriniakis et al. further defines

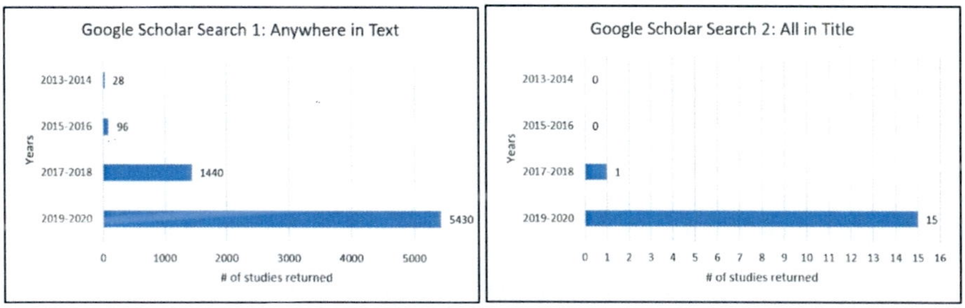

Fig. 1. Google Scholar Search Result Sets for text: "digital twin" AND ("artificial intelligence" OR "machine learning" OR "deep learning" OR "cognitive"), 2020.11.09.

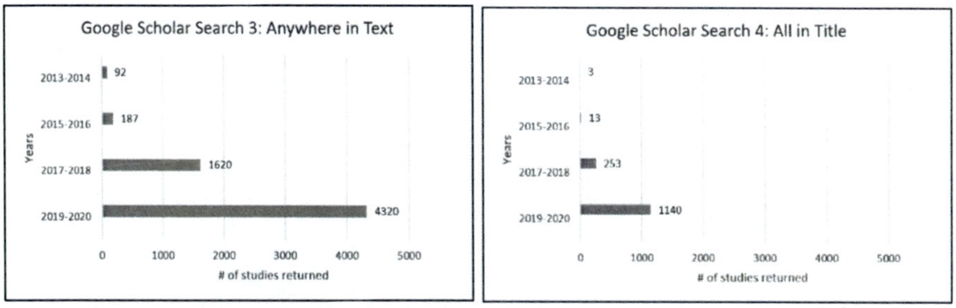

Fig. 2. Google Scholar Search Result Sets for text: digital twin–"artificial intelligence"-"machine learning"-"deep learning"-cognitive, 2020.11.09

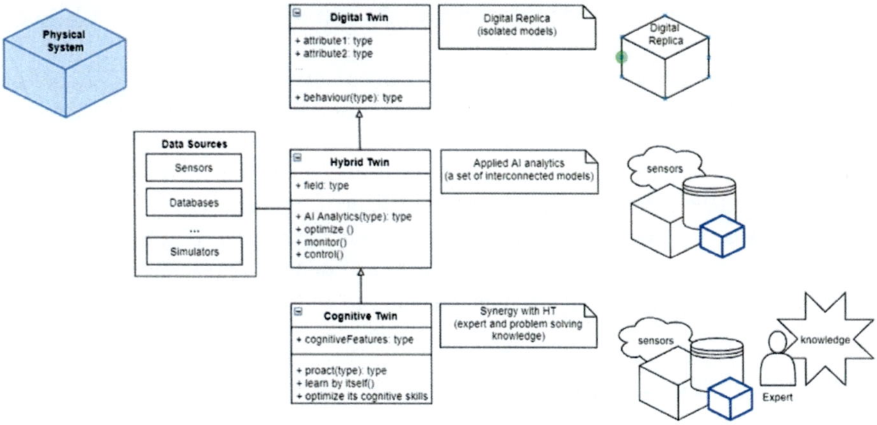

Fig. 3. Digital, hybrid and cognitive twins: a taxonomy for digital twins according to Abburu et al.

Enhanced Cognitive Twins (ECT), and introducing advanced cognitive capabilities to the Digital Twin artefact enabling decision support, aiming Digital Twins to react to inner or outer stimuli [11]. The ECT can be deployed at different hierarchical levels of the manufacturing process, in order to tolerate both horizontal and vertical interplay. The ECT is proposed in the context of process industries, where cognition is particularly important due to the continuous, non-linear, and varied nature of the respective production processes [11].

2.2 Digital Twins in the Steel Industry

Being one of the process industries, steel industry provides many use cases for the use of Cognitive Digital Twins. For efficient manufacture of high-quality steel products, digitalization of systems, instrumentation, and control technologies using computers are essential [12, 16, 35].

Effective utilization of AI and big-data for steel manufacturing processes is actively being pursued not only in Europe, but also in other countries, such as Japan [35].

Digital Twins for steel manufacturing plant can be grouped as: 1) equipment (or machine) level Digital Twin, 2) process level Digital Twins, and 3) plant level Digital Twins.

Digital Twins will include, but not limited to

- equipment level Digital Twin that control each equipment,
- sub-blocks that will enable building equipment level Digital Twin,
- connection between Digital Twins and existing control systems,
- process level Digital Twin that will control equipment within process and
- plant level Digital Twin that will control entire operations of plant [10].

There are various use cases for Digital Twins in the steel industry. Experimenting multiple ratios of carbon in steel to arrive at suitable products is one of them [16]. Digital Twins can also be used in determining the optimal amount of materials to be used.

Another important use case is in the area of maintenance. Industrial Internet of Things (IIoT), Digital Twin, and ML play relevant roles in the context of maintenance [14, 32]. New opportunities for predictive maintenance of manufacturing machines are introduced by Digital Twins [28, 30, 31]. In the steel industry, the digitalization do not only produce quality, flexibility, and productivity, but also make sure visibility of the real-time data and provide insight for better and faster decision making [6]. Energy management, water and waste management including pollution prevention and control are other utilization areas of digital twin applications in the steel industry [6].

3 The CogniTwin Project

The Cognitive Plant Through Proactive Self-Learning Hybrid Digital Twins (CogniTwin) project has received funding from the European Union's Horizon 2020 research and innovation programme under GA No.870130 [33].

The project has 13 participants from seven different countries: Finland, France, Germany, Norway, Serbia, Spain, Turkey.

As listed in Table 1, all of the six industrial partners are from the process industries. The industrial partners are composed of non-ferrous (Hydro, Elkem), steel (Sidenor, Saarstahl, Noksel), and engineering (SHI FW Energia). The technology providers are all small medium enterprises: Cybernetica, NST, Scortex and Teknopar. The research partners include SINTEF, Fraunhofer, DFKI and University of Oulu.

The CogniTwin project aims to add the cognitive elements to the existing process control systems. By doing so, the project aims enabling the control systems capability to self-organise and offer solutions to unpredicted behaviours [33]. The CogniTwin project is an ongoing project. The project started on September 1st 2019, and its duration is 36 months. CogniTwin aims to bring the industrial partners to a new level of Industry 4.0-driven operation by bringing in new data sources, integration of new and existing data, applying ML techniques to generate hybrid, self-learning and proactive systems.

The main objective of the project is to achieve improved performance in Cognitive Production plants in the SPIRE Process Industry sectors of Aluminum, Silicon (Non Ferrous), Steel and Engineering through the use of Hybrid and Cognitive Digital Twins [32]. The other objectives include:

- Industry process excellence,
- Cognitive retrofitting,
- Optimized process performance,
- Development of interoperability toolbox as a service,
- Increased European technology dominance, and
- Ensuring technology transfer.

Within the scope of CogniTwin, there are six large scale pilots. Each pilot is related to one of the associated industrial partners. TEKNOPAR [34] is the technology provider for the NOKSEL [26] pilot case. Next, the NOKSEL pilot is detailed.

4 The NOKSEL Case

NOKSEL is among the top 100 biggest companies in Turkey. The company follows API standards, and manufactures pipes in accordance with AWWA [3], BSI [7], ASTM [4], ISO [17], EN [13], and AFNOR [2]. NOKSEL provides spiral welded steel pipes for different usage areas, such as petroleum, gas, water and construction industries. In Turkey, there are two NOKSEL plants. The NOKSEL case of CogniTwin is performed in NOKSEL's İskenderun plant. At this plant, spiral welded steel pipe manufacturing machine (SWP) is used to produce steel pipes. SWP contains different components for preparing the pipe, welding and cutting. Figure 4 shows the SWP process at NOKSEL's İskenderun plant. In the SWP process, unprocessed steel sheets are used as input, and spiral welded steel pipes are produced as output. Feeding, rolling, levelling, edge milling, forming, internal welding, external welding and length cutting are parts of the SWP process.

The NOKSEL case is about cognitive digital twin powered condition monitoring in steel pipe manufacturing. The case also aims to perform predictive maintenance of the SWP machinery.

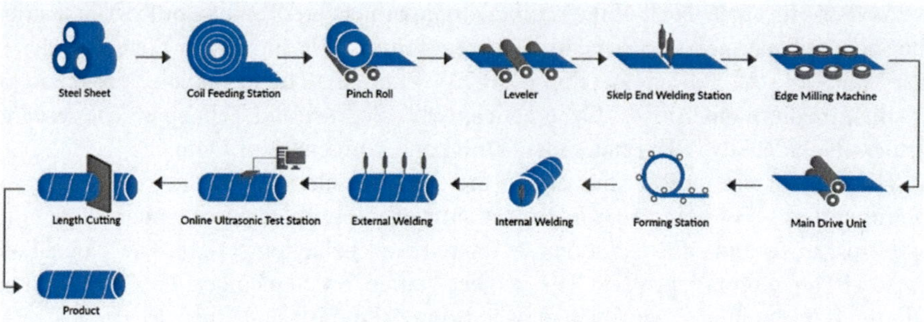

Fig. 4. SWP process at NOKSEL İskenderun Plant

4.1 Objectives

The objectives of the NOKSEL case on the SWP machinery are threefold: 1) real-time condition monitoring, 2) predictive maintenance, and 3) digital twin generation.

Due to the very large size of SWP machinery, and the high number of its components, development of a user-friendly software for SWP's real-time condition monitoring is a challenge. The condition monitoring will be realized by means of a platform. The platform will be connected with PLC, SCADA and ERP interfaces of the production plant. In the aimed platform, data gathered from different sources, such as sensors, operational data input, automation system input will be analyzed for understanding current situation of the SWP machinery. Using the digital twin technology, SWP's output will be shown to the operator. Smart components that use sensors to gather data about real-time status, working condition, or position will be connected to a cloud-based system.

By applying AI/ML/DL based analytics and computation, and using the data acquired by the platform, smart predictive maintenance is aimed via cognitive digital twin. The Digital Twin, will visualize maintenance predictions real-time, and notify personnel with current condition of the machinery, abnormalities and alarms.

Both data driven and model driven twins are aimed to be generated for the case. Both synthetic and real data collected from the platform are planned to be used for data driven twin generation. The 1st order, or the process modeling will be used to generate the model driven twin. By combining the data and the model driven twins, a hybrid digital twin is aimed to be created. Adding the cognitive elements to conduct predictive maintenance by enhancing the digital twin via AI/ML/DL components, a cognitive digital twin is targeted to be generated.

4.2 Success Criteria

For the NOKSEL pilot case to be successful, the following KPIs are targeted:

- No errors in anomaly detection (both Type 1 and Type 2),
- 10% reduction in machine downtime, and
- 10% reduction in energy consumption.

4.3 Work Accomplished

To satisfy the objectives of the case, the following work was completed in the first third of the project:

1) Open core platform implementation for real-time condition monitoring: Following the selection of sensors to be used, sensor implementations were performed by TEKNOPAR [34]. A service-oriented cloud architecture model was designed and developed to integrate smart devices, sensors, machine models and automation system information. The platform is easy to deploy and open to add new services. To enable free connection and interoperability, data entry interfaces and functionalities of the core platform were defined and platform was developed using open source technologies (mostly from Apache foundation, and open standards). Data acquired by the platform were preprocessed for condition monitoring and anomaly detection. The collected data stream is processed real-time and stored in database.

2) Big data analytics and AI services: First, selection of AI modeling tool was conducted. After comparing alternative tools, MATLAB, Simulink, [24] and Python [27] were selected to be used for AI/ML/DL studies. Using different data sets collected from the internet, different AI models were tested and compared for predictive maintenance. First feature extraction was applied, later different algorithms were trained by the data set, and the trained models were tested, using F1 score, accuracy, precision and recall values. An application to run and compare performance of ML and DL algorithms for predictive maintenance was developed (Fig. 5). For all ML algorithms selected, best performance values are presented in a graph. In addition, the algorithms are also compared with respect to certain parameters such as, error detection rate, total training time, total test time, average training time, average test time, Type I and Type II errors.

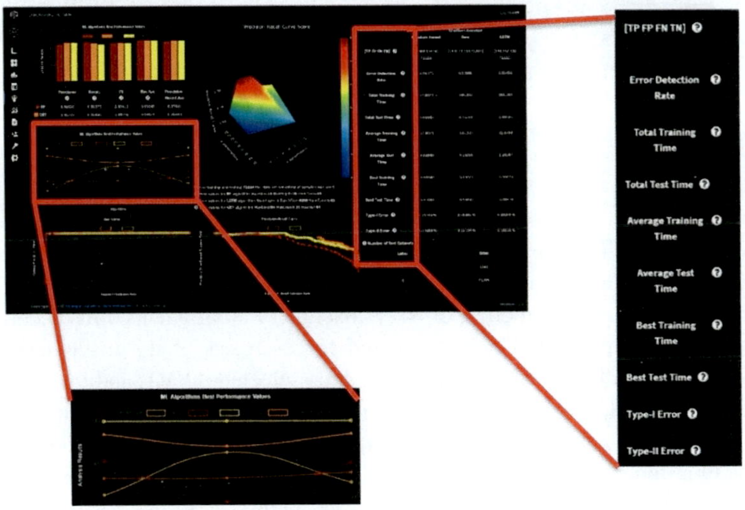

Fig. 5. AI/ML/DL algorithms application GUI

3) Development of Digital Twin visualizations: Using the core platform, hierarchical digital twins have been developed to empower facilities with virtual technologies. The development of virtual dashboards, showing condition monitoring information on Digital twins of machinery have been completed (Fig. 6). The GUIs include visual representation of the machinery, dashboard items in various forms such as gauges and line charts.

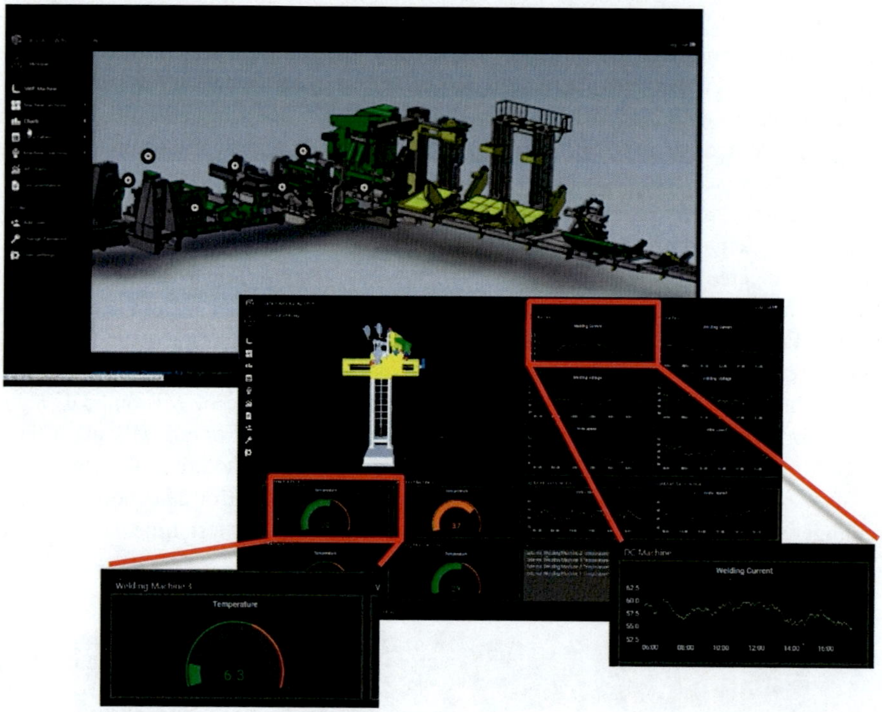

Fig. 6. Digital twin visualization GUIs

4.4 Future Work

A Cognitive Digital Twin to be used for predictive maintenance of the SWP machinery at NOKSEL's İskenderun plant will be designed and developed by TEKNOPAR. Development of 3D and AI models of SWP machinery components will be completed. The integration of the digital twins prepared for the parts of the machinery will ve performed. Using real and real-time acquired data, selected AI/ML and DL models will be further improved, trained and tested for predictive maintenance purposes of the SWP. TEKNOPAR will later take part in evaluation, verification and validation of the concepts, requirements and results for cognitive twin functionality and influence. We will define and plan experimental systems & demonstrations. Evaluation of the demonstration results will be conducted. A toolbox will be developed and tested for the other relevant pilots in the project.

The future work is not limited to the technical studies. In parallel to the technical studies in the project, exploitation and dissemination activities will be performed.

5 Conclusion

The impact and widespread use of Digital Twins together with AI/ML/DL technologies have great impacts on today's and future manufacturing world, including the steel production industry.

In this study, we briefed the details of a pilot on the design and development of cognitive digital twins related to spiral welded steel pipe production. The pilot belongs to the CogniTwin project. The first 12 months of the CogniTwin project which aims to develop self-learning proactive cognitive digital plants has been completed. During this time period, in the CogniTwin project towards developing a self-learning and proactive Cognitive Digital Twin for NOKSEL's SWP machinery, planned tasks have been successfully completed. These tasks include design and implementation of core platform real-time condition monitoring platform, big data analytics and AI services, and Digital Twin design and development for NOKSEL's SWP. Both data driven and model driven digital twins for the SWP have been developed, and as a result, a hybrid digital twin has been realized.

In the future, we will add cognitive elements to the hybrid digital twin developed for the SWP machinery, and design and develop a Cognitive Digital Twin for it.

Acknowledgement. The CogniTwin project has partially been funded by the European Union's Horizon 2020 research and innovation programme under GA No. 870130. We would like to thank the TEKNOPAR's R&D team members, and the software engineers who work for the CogniTwin project. For the use case, we thank to all NOKSEL employees who participated in the CogniTwin.

References

1. Abburu, S., Berre, A.J., Jacoby, M., Roman, D., Stojanovic, L., Stojanovic N.: COGNITWIN – Hybrid and cognitive digital twins for the process industry. In: 2020 IEEE International Conference on Engineering, Technology and Innovation (ICE/ITMC), Cardiff, United Kingdom, pp. 1–8 (2020). https://doi.org/10.1109/ice/itmc49519.2020.9198403
2. AFNOR Standards. https://www.afnor.org/en/. Accessed 29 Oct 2020
3. American Water Works Association. https://www.awwa.org/. Accessed 29 Oct 2020
4. ASTM Standards. https://www.astm.org/. Accessed 29 Oct 2020
5. Barricelli, B.R., Casiraghi, E., Fogli, D.: A survey on digital twin: definitions, characteristics, applications, and design implications. IEEE Access **7**, 167653–167671 (2019). https://doi.org/10.1109/ACCESS.2019.2953499
6. Branca, T.A., Fornai, B., Valentina, C., Murri, M.M., Streppa, E., Scröder, A.J.: The Challenge of Digitalization in the Steel Sector, Metals, Special Issue on Challenges and Prospects of Steelmaking Towards the Year 2050 (2020). https://www.mdpi.com/2075-4701/10/2/288/html. Accessed 07 Dec 2020
7. BSI. https://www.bsigroup.com/. Accessed 29 Oct 2020

8. Chakraborty, S., Adhikari, S.: Machine learning based digital twin for dynamical systems with multiple time-scales. arXiv preprint arXiv:2005.05862 (2020)
9. Deutsches Institute for Norming. https://www.din.de/de. Accessed 29 Oct 2020
10. Digital Twins for Industrial Applications: Definition, Business Values, Design Aspects, Standards and Use Cases, An Industrial Internet Consortium White Paper, Version 1.0, 08 Februrary 2020. https://www.iiconsortium.org/pdf/IIC_Digital_Twins_Industrial_Apps_White_Paper_2020-02-18.pdf. Accessed 18 Nov 2020
11. Eirinakis, P., et al.: Enhancing cognition for digital twins. In: 2020 IEEE International Conference on Engineering, Technology and Innovation (ICE/ITMC), Cardiff, United Kingdom, pp. 1–7 (2020). https://doi.org/10.1109/ice/itmc49519.2020.9198492
12. ESSA, European Steel Skills Agenda, Blueprint "New Skills Agenda Steel": Industry-driven sustainable European Steel Skills Agenda and Strategy, Company Skills Requirements and Foresight, Version 1.0, January 2020. https://www.estep.eu/assets/Uploads/ESSA-D5.2-Industry-Skills-Requirements-Version-1.pdf. Accessed 09 Nov 2020
13. EUROPEAN Standards. https://www.en-standard.eu/. Accessed 29 Oct 2020
14. Fumagalli, A.: Digital twin and machine learning in an Industry 4.0 approach to maintenance (2018). https://www.politesi.polimi.it/handle/10589/141563. Accessed 09 Nov 2020
15. Fuller, A., Fan, Z., Day, C., Barlow, C.: Digital twin: enabling technologies, challenges and open research. IEEE Access **8**, 108952–108971 (2020). https://doi.org/10.1109/ACCESS.2020.2998358
16. Hinduja, H., Kekkar, S., Chourasia, S., Chakrapani, H.B.: Industry 4.0: digital twin and its industrial applications. Int. J. Sci. Eng. Technol. Open Access J. **8**(4) (2020). ISSN 2348-4098. https://www.researchgate.net/profile/Hitesh_Hinduja/publication/343713676_Industry_40_Digital_Twin_and_its_Industrial_Applications/links/5f3ba50d299bf13404cd6750/Industry-40-Digital-Twin-and-its-Industrial-Applications.pdf. Accessed 08 Nov 2020
17. ISO. https://www.iso.org/standards.html. Accessed 29 Oct 2020
18. Jacoby, M., Uslander, T.: Digital Twin and Internet of Things - Current Standards Landscape, Applied Sciences, vol. 10. https://www.mdpi.com/2076-3417/10/18/6519. Accessed 09 Nov 2020
19. Jaensch, F., Csiszar, A., Scheifele, C., Verl, A.: Digital twins of manufacturing systems as a base for machine learning. In: 2018 25th International Conference on Mechatronics and Machine Vision in Practice (M2VIP), Stuttgart, pp. 1–6 (2018). https://doi.org/10.1109/m2vip.2018.8600844
20. Kaur, M.J., Mishra, V.P., Maheshwari, P.: The convergence of digital twin, IoT, and machine learning: transforming data into action. In: Farsi, M., Daneshkhah, A., Hosseinian-Far, A., Jahankhani, H. (eds.) Digital Twin Technologies and Smart Cities. Internet of Things (Technology, Communications and Computing). Springer, Cham (2020). https://doi.org/10.1007/978-3-030-18732-3_1
21. Kharchenko, V., Illiashenko, O., Morozova, O., Sokolov, S.: Combination of digital twin and artificial intelligence in manufacturing using industrial IoT. In: 2020 IEEE 11th International Conference on Dependable Systems, Services and Technologies (DESSERT), Kyiv, Ukraine, pp. 196–201 (2020). https://doi.org/10.1109/dessert50317.2020.9125038
22. Lee, J., Azamfar, M., Singh, J., Siahpour, S.: Integration of digital twin and deep learning in cyber-physical systems: towards smart manufacturing. IET Collaborative Intell. Manuf. **2**(1), 34–36 (2020). https://doi.org/10.1049/iet-cim.2020.0009

23. Lim, K.Y.H., Zheng, P., Chen, C.: A state-of-the-art survey of Digital Twin: techniques, engineering product lifecycle management and business innovation perspectives. J. Intell. Manuf. **31**, 1313–1337 (2020). https://www.researchgate.net/publication/337363063_A_state-of-the-art_survey_of_Digital_Twin_techniques_engineering_product_lifecycle_management_and_business_innovation_perspectives. Accessed 08 Nov 2020
24. MATLAB. https://www.mathworks.com/products/matlab.html. Accessed 29 Oct 2020
25. Minerva, R., Lee, G.M., Crespi, N.: Digital Twin in the IoT context: a survey on technical features, scenarios, and architectural models. Proc. IEEE **108**(10), 1785–1824 (2020). https://doi.org/10.1109/JPROC.2020.2998530
26. NOKSEL. http://www.noksel.com.tr/. Accessed 29 Oct 2020
27. Python. https://www.python.org/. Accessed 29 Oct 2020
28. Qiao, Q., Wang, J., Ye, L., Gao, R.X.: Digital Twin for machining tool condition prediction. Procedia CIRP **81**, 1388–1393 (2019). https://www.sciencedirect.com/science/article/pii/S2212827119306638. Accessed 09 Nov 2020
29. Rasheed, A., San, O., Kvamsdal, T.: Digital Twin: Values, Challenges and Enablers. arXiv preprint arXiv:1910.01719 (2019)
30. Riantama, R.N., Prasanto, A.D., Kurniati, N., Anggrahini, D.: Examining equipment condition monitoring for predictive maintenance, a case of typical process industry. In: Proceedings of the 5th NA International Conference on Industrial Engineering and Operations Management Detroit, Michigan, USA, 10–14 August 2020. http://www.ieomsociety.org/detroit2020/papers/690.pdf. Accessed 08 Nov 2020
31. Roy, R.B. Mishra, D., Pal, S.K., Chakravarty, T., Panda, S., Chandra, M.G., Pal, A., Misra, P., Chakravarty, D., Misra, S.: Digital twin: current scenario and a case study on a manufacturing process: Int. J. Adv. Manuf. Technol.1–24 (2020)
32. The COGNTWIN Project Proposal
33. The COGNITWIN Project. https://cordis.europa.eu/project/id/870130. Accessed 29 Oct 2020
34. TEKNOPAR. https://teknopar.com.tr/. Accessed 29 Oct 2020
35. Yoshizawa, I., Nakagawa, S.: Development of Systems, Instrumentation, and Control Technologies for Steel Manufacturing Processes and Future Prospects, Technical Review: NIPPON STEEL & SUMITOMO METAL Technical Report No. 121, March 2019. https://www.nipponsteel.com/en/tech/report/nssmc/pdf/121-02.pdf. Accessed 08 Nov 2020

TSorage: A Modern and Resilient Platform for Time Series Management at Scale

Mathieu Goeminne(✉), Nico Salamone, Mohamed Boukhebouze, and Stéphane Mouton

Centre d'Excellence en Technologies de l'Information et de la Communication, Gosselies, Belgium
{mathieu.goeminne,nico.salamone,mohamed.boukhebouze, stéphane.mouton}@cetic.be

Abstract. This paper presents an open source platform for managing time series at scale. The platform provides innovative functional and non-functional features and services that help the user in addressing current challenges in data collection, processing and storage in an industrial context. Opportunities of using this platform for developing and managing artificial intelligence systems are discussed.

Keywords: Industry · IoT · Time series · Big data · Platform

1 Introduction

Steel manufacturing companies are facing an increasingly aggressive worldwide competition. The Integrated Intelligent Manufacturing roadmap defined by the ESTEP working group [1] considers the Industrial Internet of Things (IIoT) as a key opportunity for strengthening their position or gaining competitive advantages in the race for better productivity, better product quality, and better process control.

However, the management of the velocity and volume of time series data provided by these IIoT represents a challenge in which the value of traditional data management solutions for the industry tends to decrease, so that the steel industry sector is looking for tools and approaches able to face scalability, availability, reusability, integration, and pricing issues, among others.

In order to meet the industry needs, we propose a new innovative solution for time series management at scale called TSorage [2]. This solution is a modular and open source platform which enables the collection, processing, storage, and visualization of time series data streams generated by industrial IoT like detectors, probes, and other sensors in the broadest sense of the term.

The remaining of this paper is structured as follows. Section 2 reviews related works. Section 3 presents the main features of the proposed solution. The constitutive elements of the platform architecture are described in Sect. 4. Section 5 presents an industrial case study for the platform. Finally, Sect. 6 concludes by summarizing the current state of development of TSorage and discussing future extensions.

V. Colla and C. Pietrosanti (Eds.): ESTEP 2020, AISC 1338, pp. 144–150, 2021.
https://doi.org/10.1007/978-3-030-69367-1_12

2 Existing Work

In the literature, several time series management solutions have been proposed by different editors. A comparison of some of these solutions is provided by Acreman *et al.* [3].

A first group of the time series management tools concerns commercial solutions which provide holistic tools that are often used by industrial stakeholders. Examples of such solutions are InfluxDB [4], Aveva Historian [5], Aspentech IP21 [6], PI Historian [7] and ThingsBoard [8]. However, these solutions do not ensure high performance processing and scalable storage of time series data (hundreds of thousands of points/seconds). In addition, most of these data historians conducts to a vendor lock-in, which makes industrial stakeholders dependent on the proposed ecosystem solution and unable to use other legacies or complementary tools without substantial switching costs.

A second group of time series software concerns new generation solutions which rely on the big data technologies to provide scalable and high-performance tools like Datadog [9] and Spotify Heroic [10]. Still, these tools tend to focus on specific activity domains, or they are on an early development stage and do not propose yet a packaged solution to help to effortless deployment and data integration. Moreover, these solutions are not designed to deal with different time series data types and tags (in this context, a tag is an arbitrary key associated with a value) that the heterogeneous IIoT could have.

In the next section, we present our innovative time series management platform (TSorage) can deal this these mentioned challenges.

3 TSorage

Using the knowledge acquired in previous industrial research projects, we identified needs and concerns frequently shared by industrial stakeholders about the development of IIoT solutions. We decided to implement an IoT platform that puts the focus on fitting these issues. The result is a collection of integrated services for managing time series of any type with millisecond resolution timestamp and as few technical constraints as possible. Among others, TSorage is designed with the following principles in mind.

Availability and Scalability. Due to its distributed and decentralized nature, TSorage capabilities can be extended by adding more resources on a cluster of servers made of commodity hardware. Hardware redundancy ensures an interrupted service can be restarted on another server. Collected time series values are replicated on multiple servers, so that the failure of a server, a rack, or even an entire data center, does not compromise data integrity or availability. While it is generally considered that the value of time series datapoints quickly decreases over time from an operational point of view, the long term storage of high resolution time series is valuable from a machine learning (or, more generally, data analytics) perspective, since the analysis of subtle and fast changes in the considered processes may improve predictions or data understanding

[11–13]. In the manufacturing domain, an anectodical example would be a pump that starts and stops multiple times per minute, highlighting an unfrequently observed and presumably undesirable behavior. With a temporal resolution of one minute or higher, such a behavior would not be easily discernible. This points to a need for highly scalable solutions for data processing and storage.

Vendor Lock-in Mitigation. TSorage is made of loosely coupled open source software components. Consequently, they can be easily replaced with alternatives as long as the latter provide the same service. Inter-component communications are provided by widely adopted technologies such as HTTP and MQTT. It fosters the use of TSorage as a technical platform for developing business- and process-specific applications.

Extensible Type Support. Most sensors express captured signals as scalar quantities. However, as sensors complexity increases, supporting other data types, such as geographic positions or complex status reports, becomes even more interesting. TSorage provides a built-in support for the commonly used data types, and is designed to be easily extended by composing basic data types with lists, sets, and dictionaries.

Effortless Data Integration. A long and painstaking procedure for extending the data source coverage would stifle any innovation. TSorage's approach consists in ingesting any submitted data point first, and administering it in a second time.

Tag-based Time Series Management. TSorage implements the Metrics 2.0 data model [14]. A system of tags is used for making sense of the time series values, whilst a single sensor name is generally used for this purpose. Tags are used to indicate the intrinsic properties of a sensor, as well as additional information about individual measurements (for instance, their quality). This approach facilitates the automatic exploration of the historical dataset since data can be filtered and broken down according to an arbitrary tag.

4 Architecture

In order to provide the features proposed in Sect. 3, TSorage is proposed as a set of easily runnable and scalable **services**. These services are organized into clearly segregated **layers**, as shown in Fig. 1. These layers are detailed below.

Gateway Layer. Data points are collected by the Gateway layer, that supports commonly used industrial communication technologies, including Modbus, OPC-UA, and MQTT. Remote databases can also be consulted for collecting historical datasets. Collected time series values are temporarily stored into a buffer, and eventually transmitted to the Ingestion layer by using HTTP or MQTT protocols.

Ingestion Layer. This layer is responsible for accepting incoming time series values. Its checks the conformity of the received messages with respect to predefined schemes and rejects unauthorized messages. Messages can be transmitted through an HTTP entry point or a MQTT broker, and can be encoded either in JSON or in Protobuf. Accepted messages are pushed to an Apache Kafka [15] topic, that acts as a message queue for the internal TSorage processing services.

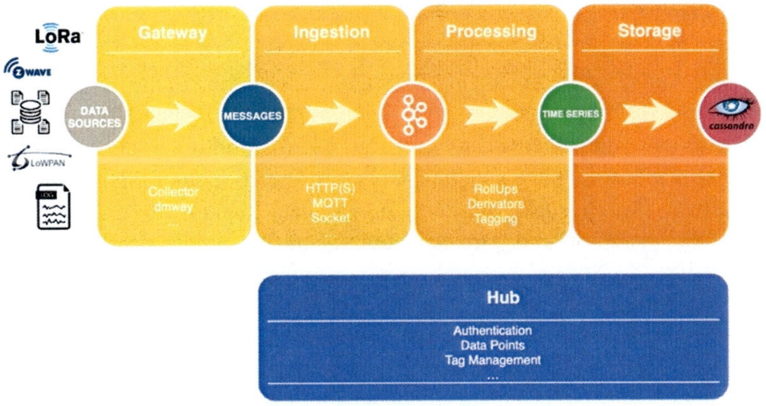

Fig. 1. Architecture overview

Processing Layer. This layer is implemented as a set of stream processing applications [16] that handle Kafka message flows. It performs data rollups and data transformations in real-time. **Data rollups** consist in aggregating the values over time, in such a way high-level analyses, that cover long time periods, can be carried out more easily. They also make cross-time series analyses easier to perform since their values are chronologically aligned. Data transformations consist in modifying the time series values, names, and tags in real-time according to business-specific functions in order to enrich the processed information.

Storage Layer. Apache Cassandra [17] is used for permanently storing time series values. This distributed and decentralized database provides a linear scalability, whilst its automatic cross-site data replication feature allows for remaining operational and recovering after the failure of a node or even an entire data center. We designed a **data model** that complies with the nature of the manipulated data and typical queries, that relate to a specific time series and a specific time range.

Beside the time series values, the database also stores metadata that helps in reaching the appropriate partitions more efficiently. Among others, a list of all the partitions associated with a given tagset of a particular metric is kept up to date in order to speed up queries responses.

Hub Layer. Time series are managed and accessed through a set of microservices that implement various services, including tagset administration, user authentication, and data points querying. This layer offers a partial implementation of PromQL [18], the query language proposed by Prometheus for selecting, aggregating, and retrieving time series data. That means visualization and analytics tools supporting this language, such as Grafana [19], integrate seamlessly with TSorage.

5 Case Study

As a first case study, we are evaluating TSorage in the context of the industrial research project ARTEMTEC [20], that aims to provide modern and advanced analytical tools from big data and augmented reality technologies for improving and optimizing the maintenance of industrial sites geographically distributed around the globe.

Our partner, Safran Aero Boosters, brings data streams from various sensors measuring the activity of aerospace test equipment. These data streams are recorded and processed with TSorage for anomaly detection purpose. The expected output of this project is the development of innovative services that meet the expectations of Safran AeroBooster's market.

Among other things, we developed a simple tool that helps the user (typically, a production operator) in specifying a list of sensors that measure interrelated phenomena. The user can also specify an anomaly detection algorithm among those we currently support, as well as a time range covered by the historical dataset. Based on these parameters, a service of the Hub layer uses PyOD [21], an open source Python library, to train a machine learning model. The model is transmitted to the Processing layer for real time scoring. The model is then considered as a virtual sensor: whenever a value that relates to one of the previously specified sensors is processed, the model calculates a value representing a *degree of abnormality*. This new value is then submitted to the Ingestion layer as a data point belonging to a time series associated with the model. Thanks to this approach, the degree of abnormality reported by the model can be visualized and further processed as any other time series.

We are considering the possibility of adding a feedback mechanism enabling users to specify whether a situation should be considered as normal or abnormal. We expect this additional information to be useful for refining detection thresholds and, consequently, for reducing the number of false positive and false negative anomaly alerts.

6 Conclusion and Future Work

The development of TSorage platform is part of our roadmap for industry 4.0 in which the ambition is to develop a data management tools portfolio that proposes modern vision of data management and complies with the paradigm shift introduced by the Industrial Internet of Things. TSorage is a modular, scalable, and resilient platform based on open source components for collecting, processing, and storing industrial time series values. A first functional version is freely available [2] and is being evaluated as part of an industrial research project. Based on our early trials, we plan to extend its functional coverage in the near future. We consider integrating ElasticSearch [22] for supporting fuzzy queries on time series names and tags.

We also plan to add a **digital twin engine** [23, 24] that compares the formal description of process dynamics with the actual operational conditions of their executions. This allows to assess to which extent a given procedure has been followed, and to determine which operational procedure brought to best process executions.

We intend to use digital twins for **simulating** the impact of hypothetical changes on a process execution. Based on "what if" scenarios, reinforcement learning algorithms

[25] will be used for preparing control systems in simulated environments, so that evidences of their reliability can be provided before exploiting them in the actual plant.

Notably, TSorage is deployed using FADI [26], a framework for automated deployment and configuration of complex software stacks developed at our research center. FADI is already used in several projects to deploy cloud native platform for Big Data based on mature open source tools. In addition, in parallel to TSorage, our research center maintains and extends an open source realistic time series generator, called TSimulus [27].

Acknowledgement. This work was carried out as part of the ARTEMTEC project, supported by Wallonia (Walloon Region) under grant agreement N°7904.

References

1. Brummayer, M., Chust, R., Colla, V., Gailly, E., Delsing, J., Kämper, F., Kuiper, G., Krauth, P.N., Mathis, G., Mouton, S., Peters, H., Pietrosanti, C., Sanfilippo, F., Winter, G.: Roadmap: Integrated intelligent manufacturing (i2m). ESTEP Working Group (2016)
2. CETIC asbl, TSorage documentation. https://doc.tsorage.io. Accessed 10 Nov 2020
3. Acreman, S., Gies, H., Brazil, B.: Open source timeseries DB comparison. https://docs.google.com/spreadsheets/d/1sMQe9oOKhMhIVw9WmuCEWdPtAoccJ4a-IuZv4fXDHxM/pubhtml. Accessed 10 Nov 2020
4. InfluxData, InfluxDB. https://influxdata.com. Accessed 10 Nov 2020
5. Aveva, Aveva Historian. https://www.aveva.com/en/products/historian/. Accessed 10 Nov 2020
6. Aspentech, InfoPlus21. https://www.aspentech.com/en/products/msc/aspen-infoplus21. Accessed 10 Nov 2020
7. Automated Results, PI Historian. http://www.automatedresults.com/PI/pi-historian.aspx. Accessed 10 Nov 2020
8. Thingsboard Inc, Thingsboard. https://thingsboard.io. Accessed 10 Nov 2020
9. DataDog Inc., DataDog. https://www.datadoghq.com. Accessed 10 Nov 2020
10. Spotify, Heroic. https://spotify.github.io/heroic/. Accessed 10 Nov 2020
11. Van Duynhoven, A., Dragicevic, S.: Analyzing the effects of temporal resolution and classification confidence for modeling land cover change with long short-term memory networks. Remote Sens. **11**, 2784 (2019)
12. Machine Metrics, Unlocking the Power of High-Frequency Control Data. https://www.machinemetrics.com/blog/unlocking-the-power-of-high-frequency-control-data. Accessed 03 Dec 2020
13. Huchtkoetter, J., Reinhardt, A.: On the impact of temporal data resolution on the accuracy of non-intrusive load monitoring. In: Proceedings of the 7th ACM International Conference on Systems for Energy-Efficient Buildings, Cities, and Transportation, BuildSys 2020, pp. 270–273, New York, NY, USA (2020). Association for Computing Machinery
14. Metrics 2.0 Community, Metrics 2.0. http://metrics20.org. Accessed 10 Nov 2020
15. Apache Foundation, Apache Kafka. https://kafka.apache.org. Accessed 10 Nov 2020
16. Theorin, A., Bengtsson, K., Provost, J., Lieder, M., Johnsson, C., Lundholm, T., Lennartson, B.: An event-driven manufacturing information system architecture for industry 4.0. Int. J. Product. Res. **55**(5), 1297–1311 (2017)

17. Apache Foundation, Apache Cassandra. https://cassandra.apache.org. Accessed 10 Nov 2020
18. Prometheus Authors, Querying Prometheus. https://prometheus.io/docs/prometheus/latest/querying/basics/. Accessed 10 Nov 2020
19. Grafana Labs, Grafana. https://grafana.com. Accessed 10 Nov 2020
20. CETIC asbl, ARTEMTEC. https://www.cetic.be/ARTEMTEC-en. Accessed 10 Nov 2020
21. Zhao, Y., Nasrullah, Z., Li, Z.: PyOD: a python toolbox for scalable outlier detection. J. Mach. Learn. Res. **20**(96), 1–7 (2019)
22. ElasticSearch B.V., ElasticSearch. https://www.elastic.co. Accessed 10 Nov 2020
23. Rauch, L., Pietrzyk, M.: Digital twins as a modern approach to design of industrial processes. J. Mach. Eng. 19, 86–97 (2019)
24. Tao, F., Zhang, H., Liu, A., Nee, A.Y.C.: Digital twin in industry: state-of-the-art. IEEE Trans. Ind. Inform. **15**(4), 2405–2415 (2019)
25. Kaelbling, L.P., Littman, M.L., Moore, A.W.: Reinforcement learning: a survey. J. Artif. Intell. Res. **4**(1), 237–285 (1996)
26. CETIC asbl, FADI. https://github.com/cetic/fadi. Accessed 10 Nov 2020
27. CETIC asbl, TSimulus. https://github.com/cetic/TSimulus. Accessed 10 Nov 2020

Author Index

V. Colla and C. Pietrosanti (Eds.): ESTEP 2020, AISC 1338, pp. 151–152, 2021.
https://doi.org/10.1007/978-3-030-69367-1